Edward Curtis

Manual of General Medicinal Technology Including Prescription-Writing

Third Edition

Edward Curtis

Manual of General Medicinal Technology Including Prescription-Writing
Third Edition

ISBN/EAN: 9783337779047

Printed in Europe, USA, Canada, Australia, Japan

Cover: Foto ©berggeist007 / pixelio.de

MANUAL

OF

General Medicinal Technology

INCLUDING

PRESCRIPTION-WRITING

BY

EDWARD CURTIS, A.M., M.D.

Emeritus Professor of Materia Medica and Therapeutics, College of Physicians and Surgeons, Medical Department of Columbia College, in the City of New York

THIRD EDITION

Conformed to the U. S. Pharmacopœia of 1890

NEW YORK

WILLIAM WOOD & COMPANY

1895

PRESS OF
STETTINER, LAMBERT & CO.,
22, 24 & 26 READE ST.,
NEW YORK.

PREFACE.

The topics belonging to medicinal technology, especially *prescription-writing*, find such scant treatment in the text-books of materia medica, that the author has been tempted to write out his lectures on these subjects, as formerly delivered in his college-course on *Materia Medica and Therapeutics.* The lecture style has been retained as the best for actual teaching. The technicalities of prescription-writing are discussed at especial length, and are made to include so much of the elements of Latin as are necessary to the art; a feature which the author's experience in teaching makes him believe will be welcome to many undergraduate readers. The technology of prescribing according to the metric system is also treated of in full, as the importance of the subject deserves. The spellings *gramme* and *centimeter*, respectively, are adopted without prejudice, simply because they are the spellings of the U. S. Pharmacopœia.

E. C.

27, WASHINGTON PLACE, NEW YORK,
August 23, 1889.

PREFACE TO THE THIRD EDITION.

IN this edition the text has been made to conform to the U. S. Pharmacopœia of 1890, whose authority became standard January 1, 1894.

E. C.

27, WASHINGTON PLACE, NEW YORK,
March 26, 1894.

CONTENTS.

PART I.

TECHNOLOGY OF MEDICINES.

CHAPTER I.

CHAPTER II.

CHAPTER III.

CHAPTER V.

PART II.

TECHNOLOGY OF MEDICATING.

CHAPTER I.

GENERAL MEDICINAL TECHNOLOGY.

INTRODUCTION.

The subject of general medicinal technology naturally divides itself into two parts, namely—first, the technology of *medicines*, and, secondly, the technology of *medicating*. To the former of these divisions belong the topics, severally, concerning medicines, of the *authority* for the same, the technicalities of their *naming* and of their *forms*, the methods of *determining quantities* of them, and the art of their *prescription*; while in the latter are included the two subjects of the different methods of *applying* medicines, and of the regulation of *doses*. These various topics will be treated of in the order named.

PART I.

TECHNOLOGY OF MEDICINES.

CHAPTER I.

THE AUTHORITY FOR MEDICINES.

By the phrase "the authority for medicines," we mean the authority which names and defines drugs, establishes standards of strength and of purity, and issues formulæ for pharmaceutical preparations. Such authority, in the case of nations of imperial government, is commonly vested in the government itself, but in the United States, following the genius of our institutions, the matter is one of voluntary submission by the two professions, respectively, of medicine and of pharmacy to a self-devised arrangement. And that arrangement is as follows: Every ten years representative delegates from certain organized bodies within the two professions named meet in convention, and, after fixing matters of general policy, appoint a committee from their numbers to prepare and publish the detailed authorizations in the shape of a book, entitled, as such books generally are, *The Pharmacopœia.* After arranging for the calling of a successor, the Convention then dissolves and its residual committee proceeds to its task. In due season the Pharmacopœia appears, and its provisions are tacitly accepted as authoritative by the physicians and pharmacists of the country, until the lapse of another decade brings about a new revision. The "Pharmacopœia of the United States" was thus first established by Convention in 1820, and the revision now in force is the seventh thereafter, being the outcome of the Convention of 1890. This revision will stand until the Convention of 1900 shall take the matter in hand, but by a special provision its authors are "authorized and directed to publish a supplement at the end of five years, if necessary." The Convention of 1900 will come into being in accordance with the following resolution of the Convention of 1890:—

"The President of this Convention shall, on or about the first day of May, 1899, issue a notice requesting the several bodies represented in the Conventions of 1880 and 1890, and also such other incorporated State Medical and Pharmaceutical Associations, and incorporated Colleges of Medicine and Pharmacy, as shall have been in continuous operation for at least five years immediately preceding, to elect a number of delegates, not exceeding three, and the Surgeon-General of the Army, the Surgeon-General of the Navy, and the Surgeon-General of the Marine Hospital Service, to appoint, each, not exceeding three medical officers, to attend a General Convention for the Revision of the Pharmacopœia of the United States, to be held in Washington, D. C., on the first Wednesday of May, 1900."

CHAPTER II.

THE NAMING OF MEDICINES.

THE naming of medicines is one of the functions of a pharmacopœia. In the United States, therefore, we follow the nomenclature established by our own Pharmacopœia. This nomenclature, as usual, embraces titles in English and in Latin, and these, following American habit, are, with us, designed to be as *short* as is consistent with proper identification. For purposes of naming, medicines divide into three categories, viz., *proximate principles*, *crude organic drugs*, and *pharmaceutical preparations*, and in our pharmacopœial nomenclature of these categories the following points are to be noted: In the case of proximate principles, the nomenclature of the science of chemistry is, in a general way, made the basis of the pharmacopœial titles, and in the revision of the Pharmacopœia now in force there is adopted for the first time, in the English titles of compounds, the modern chemical custom of putting foremost the name of the basylous or metallic element. So the old-time official title *Chloride of*

Sodium becomes now *Sodium Chloride*, and the title *Nitrate of Silver* is changed into *Silver Nitrate.* In the Latin titles the type followed is exemplified by the title *Sodii Chloridum*—a phrase which is a proper rendering of either of the styles of English title for the substance in question, recited above. In a few categories of compounds the action of the Pharmacopœia is peculiar, as follows: 1. In the case of salts of the alkali-metals of different grades, which, in modern chemical parlance, are of *normal* and *acid* composition, respectively, the pharmacopœial nomenclature is that of the chemistry of old, according to which a *normal* salt, as, for instance, normal potassium carbonate, is styled, simply, *Potassium Carbonate*, and the corresponding acid salt *Potassium Bicarbonate.* 2. In the case, again, of metallic compounds which, although containing identical radicals, present the same in different proportion, the modern chemical scheme of turning the name of the metal into an adjective with endings in *-ous* and *-ic*, respectively, is, in some cases (iron and mercury compounds), followed in the English pharmacopœial titles, but never in the Latin. In the Latin titles the metallic component is always expressed by the use of the substantive name of the metal, set in the genitive case, necessary contradistinction between two grades of compounds of identical radicals being expressed by some device peculiar to the case, according to practical considerations of well-established understanding on the one hand, or safety on the other. 3. With *compounds having a well-known non scientific name*, the same is often adopted in preference to the chemical title. Hence, in the place of *potassio-aluminium sulphate*, *ethyl oxide*, *trichloromethane*, or *phenol*, we have as the respective official titles for these substances the more familiar names, *Alum*, *Ether*, *Chloroform*, and *Carbolic Acid*.

Concerning, next, the nomenclature of *crude organic drugs*, the principles followed are thus succinctly stated in the preface to the Pharmacopœia itself:[1] "1. The official Latin title of a vegetable drug is to be the botanical genus-name. A few titles were excepted from this rule, being those of old and well-known drugs, as *Belladonna*, *Frangula*, *Ipecacuanha*, *Pulsatilla*, *Senna*, *Stramonium*, etc. 2. The official Latin title, selected according to the preceding rule, is to denote, or stand for, the *part* of the plant directed to be used, provided that only *one* part of the plant is official. Examples: *Aconitum*, to stand for Aconite Root; *Conium*, for Conium Seed;

---

[1] Revision of 1880.

Hyoscyamus, for Hyoscyamus Leaves, etc. But if more than *one* part is in use, the part is to be specially mentioned in the title. Examples: *Belladonnæ Folia; Belladonnæ Radix; Stramonii Folia; Stramonii Semen.* 3. The official English titles are to be the scientific, botanical (genus or species) names rather than the vernacular names; except in the case of those drugs where the vernacular names are derived from and [are] still almost identical with the scientific names, or where long custom has sanctioned some other name."

In Latin titles, where two nouns occur, or a noun and an adjective, the Latin idiom of order of sequence of the words is followed. Hence the titles *Leaves of Belladonna* and *Purified Aloes* are rendered, respectively, *Belladonnæ Folia* and *Aloë purificata.*

Lastly, the names of *pharmaceutical preparations* are arbitrary, and will present themselves seriatim when we discuss the forms of such preparations. The only general point to note in this place is that, in latinizing, the Latin idiom just cited is *not* followed as regards the position of the genus-name of the preparation. Thus the title *Tincture of Opium* is not rendered *Opii Tinctura*, as Latin usage would naturally have it, but *Tinctura Opii.* Adjectives, however, when occurring, are put in their proper Latin place, following their noun, as exemplified in the titles *Tinctura Opii Deodorati*, "Tincture of Deodorized Opium"; *Extractum Gentianæ Fluidum*, "Fluid Extract of Gentian."

CHAPTER III.

FORMS OF MEDICINES.

The *forms of medicines* next require consideration, and our study here must be precise, for the products of pharmacy have class-titles and class-peculiarities which must be thoroughly understood by the prescriber. These products are most conveniently grouped for study into those for general and those for special application. Of the former, in the case of solid drugs, the simplest form is the crude drug reduced to *powder*. Pulverization is an obvious prerequisite for the majority of applications of drugs, and where, in a prescription, the powdered condition of an ingredient is a plain necessity, the pharmacist, in compounding, uses the powder without the physician being obliged specifically to order the same. But also certain powders, simple and compound, are among the prescriptions of the Pharmacopœia, constituting thus a kind of preparation under the title PU'LVIS, *Powder*. Limitations in the use of the powder as a form of medicine are that deliquescent and oily substances do not keep well in such form, and that corrosive, bad-tasting, or bulky medicaments cannot well or wisely be so administered. But where these conditions do not obtain, the powder is a convenient form, since the doses can be put up in separate paper packages, easily carried about and easily administered. According to the nature of the substance the powder can be taken dry, or in solution, or stirred into molasses, honey, or preserve, or enclosed within a mass of some pulpy substance, such as apple-scrapings, or, more elegantly, for older patients, encased in the *capsules* or *wafers* sold by druggists for such purpose. A "capsule" is a cylindrical chamber of jujube paste or similar material, made in two pieces, whereof the one fits into the other with a telescope-joint. Capsules are of various sizes, the largest capable of holding three, four, or more grains of a dry vegetable powder, or, if moistened so as to pack closer, as much even as ten grains. The charged and closed capsule, it is needless to say, is to be swallowed bodily like a pill. "Wafers" are of two forms. In the one, two watch-glass shaped bodies, with the powder interposed, are joined by moistening the edges and pressing, and then the whole, rendered soft and slippery by the dipping for an instant into water, is swallowed as a bolus. In the other form the wafer is simply a large, thin, flat wafer proper, which, made limp by moistening, is wrapped about the powder, and the bolus at once swallowed. By the use of these capsules and wafers taste is wholly concealed, and bad-tasting medicines can, by their respective means, be prescribed in powder without objection. In the case of the capsule, furthermore, there is the special advantage that the pharmacist can be ordered to deliver the doses already encapsuled, a convenience both for the matter of carrying and of taking.

In the case of vegetable drugs, which contain, as of course such substances do, all sorts of inert matters, it is a plain desideratum to get the active constituents more or less perfectly isolated. The simplest treatment in this direction, designed for the yielding of a solid product, is to evaporate to a more or less solid consistence a solution containing the desired principles. Such solution may be the natural juices of the plant mechanically expressed, or one artificially obtained by treating the drug with some appropriate solvent. In either case the product derived as described is entitled EXTRA'CTUM, *Extract.*

Extracts vary extremely in consistence. Some are hard enough to undergo pulverization, and such are commonly furnished for use in the condition of powder. A considerable number, on the other hand, are quite soft, and also are of just the proper degree of stickiness for making well into pills. Extracts prepared by the method of expression are examples of the soft variety. What extracts are hard and what soft must be learned by rote, since the pharmacopœial nomenclature, in the case of extracts, does not take cognizance of the consistence of the preparation. A special class of extracts is constituted by the products obtained, by extracting with ether, from drugs containing a conjoint volatile oil and resin. A product of such kind is designated, in the Pharmacopœia, by the special title OLEORESI'NA, *Oleoresin*. Oleoresins, however, although in pharmacopœial nomenclature a distinct class of preparations, are yet extracts, since, in their making, the original solvent is gotten rid of by evaporation, as in the case of extracts, specifically so-called. Oleoresins are bodies of viscid, semifluid consistence, and are commonly prescribed, for internal giving, in capsule or in mixture.

Extracts form a tolerably numerous class of preparations, and are valuable because of their concentration, but yet are inferior to the chemically isolated active principles with which they compete, because of occasional uncertainty of strength. In the case, indeed, of extracts made by expressing the juice of a fresh plant, variability of strength is frequent, so much so that these particular extracts are distinctly unadvisable in cases where accurate dosage is necessary. Recognizing this fact, the present Pharmacopœia avoids this method of preparation in all cases of extracts, except in the single instance of the inconsequential *extract of taraxacum.*

Of *fluid* pharmaceutical products of general avail there are quite a number, the technical class-names of which are determined partly by the nature of the fluid basis, and partly by other considerations. The broadest division of these categories is into preparations, respectively, where the active ingredient is in *mere mechanical suspension* in a fluid, and, on the other hand, in *true solution.* In the former of these categories, if a *solid* (in powder, of course) is in suspension, the product is MISTU'RA, *Mixture;* but yet this title is not confined to such preparations, but applies, generically, wherever a fluid is a literal *mixture* of diverse forms, whether including a solid in suspension or not. A small number of mixtures are official. If a *fixed oil, fat,* or *resin* be in suspension in a watery menstruum, the product is EMU'LSUM, *Emulsion. Milk* is a natural example of an emulsion. Oils in emulsion are better borne by the stomach and better assimilated than when swallowed clear, and in such form also the disagreeable taste of an oil can be more easily disguised.

Mixtures and emulsions are forms which extemporaneous prescriptions very commonly take, and in devising them the following points must be observed: First, as regards *mixtures* (using the word now in the restricted sense of a solid, in powder, held in suspension in a more or less viscid vehicle), the main points are these: the *viscidity* is generally gotten by using mucilage of some kind, or a syrup or glycerin, letting the viscid constituent form from one-fourth to one-half the bulk of the mixture—the heavier and the greater the amount of powder to be suspended, the higher to be the percentage of the viscid addition. Next, the suspended matter should not be intrinsically too heavy—such things as heavy metallic compounds like calomel or the salts of bismuth being best given in some other way than in mixture. Nor should the substance in suspension, even if light, be in too large proportion. Natural vegetable powders should not be prescribed to exceed twenty per cent. of the weight of the mixture, and powdered extracts not to exceed five per cent. As regards the prescription of *emulsions*, the emulsifying agents in commonest use are, for fixed oils, *mucilages*, generally those of acacia and of tragacanth, respectively, and *yolk of egg*. If the latter substance be selected, not more of the emulsion should be ordered than is likely soon to be consumed, since it does not keep well. In any case, the oil is first to be thoroughly emulsified by rubbing with the undiluted viscid agent, and then, but not till then, the desired dilution with water, or with syrup and water, effected. The proportion of emulsifier to oil ranges from that of one to four to that of one to two. In diluting an emulsion, *saline solutions* and *alcohol*, except in small proportion, must be avoided, else a precipitate will occur. *Volatile oils* are required in emulsion only in small proportion, their presence in a mixture being generally for flavoring purpose only. The emulsion in this case is made by rubbing the oil with any solid that may be intended as an ingredient of the mixture, or, if there be none such, by rubbing with sugar, or even with strong syrup or with glycerin. In the exceptional case where considerable of a volatile oil is required in emulsion, as where a goodly dose of oil of turpentine is desired in such form, an excellent plan is to mix the same with at least an equal volume, or with double the volume, of a bland fixed oil, and then emulsify the mixture of the two substances. Olive oil, sweet-almond oil, or oil of sesamum are fixed oils available for this purpose. It is perfectly possible, however, to emulsify volatile oils directly, by the use of mucilage or of yolk of egg, as in the case of fixed oils. For the emulsifying of resins, suspension is effected, in the case of pure resins, by trituration in mucilage, and in the case of gum-resins—natural mixtures of a gum and a resin—by the simple addition of water, the gum in the gum-resin dissolving to form a mucilage in which the particles of the resin then remain in suspension.

From mixtures and emulsions, where a material is in suspension in a fluid, undissolved, we pass to forms of preparations where the medicinal thing is in actual solution in some particular fluid. And the available fluids for such purpose being several, we have a number of technically different forms of medicines, all of which consist of a thing or a number of things in solution. Such preparations are as follows:

The simple title LI'QUOR, *Solution*, is official and applies to all heterogeneous solutions that do not belong to some technically special class. All of the pharmacopœial "solutions" are solutions, in a watery basis, of chemicals—mostly salts. Such solutions partake, of course, of the medicinal qualities of the dissolved thing, and present, therefore, no class-features for comment in this place. Solution in water, or in a watery mixture, is, naturally, a common form in which extemporaneously to prescribe aqueously soluble substances, and, in resorting to such form, two poi ts only need to be remembered. These are, first, that many salts which keep indefinitely in the dry condition, may yet spoil readily in aqueous o -tion. In such case, therefore, small quantit es only of the solution should be made at a time, or else some preservative should be added. Notable examples of salts that undergo change in solution are salts of the *alkaloids*, generally, except the cinchona alkaloids; and, again *generally*, though generally only, salts of any base with the so-called "organic acids," citric, tartaric, acetic, and lactic. Even metallic citrates, etc., fall into the category, and the familiar salts *tartar emetic*, and the *citrates* and *tartrates of iron*, must, therefore, not be counted upon to survive in aqueous solution beyond a very short time. The second point to regard in the prescribing of extemporaneous solutions is the very obvious one of the degree of solubility of the salt to be dissolved. This matter of the solubility of pharmacopœial chemicals in water and alcohol was thoroughly re-tested by the committee who prepared the revision of our Pharmacopœia of 1880, and the results appear in a table in the Pharmacopœia, which is here reproduced.[1]

---

[1] See Appendix.

Contradistinguished from "Solution" is A'QUA, *Water.* This title came into being to express a *distilled water*, *i.e.*, a water medicated by distillation from some crude drug, such as an herb. By the nature of the case such preparation must be a *watery* solution of a *volatile* thing; and, indeed, the distilled waters of former days were mainly waters containing *volatile oils*, obtained by distillation from aromatic herbs. At present the class "waters" still mainly consists of watery solutions of volatile oils, although these solutions, with two exceptions (the waters, respectively, of orange flowers and of rose), are now made by direct solution of the oil itself, instead of by distillation. These aromatic waters are feeble in strength, because of the slight solubility of volatile oils in water; they are all selected from agreeably flavored oils, and are intended rather as pleasantly aromatized fluid vehicles for extemporaneous solutions or mixtures than as medicines proper. Dose—if they can be said to have an exact dose—about a tablespoonful. The aromatic waters are those, respectively, of *bitter almond*, *anise*, *orange flowers*, *cinnamon*, *fennel*, *peppermint*, *spearmint*, and *rose*. In addition the class includes a water, severally, of *camphor*, of *creosote*, of *ammonia* (two grades of strength), of *chlorine*, of *chloroform*, and of *hydrogen dioxide*.

In the case of the last-named preparation, although the Latin title is *Aqua* Hydrogenii Dioxidi, the English title is *Solution* of Hydrogen Dioxide. With this exception the rule obtains that all simple aqueous solutions of *volatile* things make technically *waters* and not *solutions*.

A distinct small class of preparations is afforded by watery solutions of *gums*—title MUCILA′GO, *Mucilage*. Four such mucilages are official, available, medicinally, as bland demulcents, or, pharmaceutically, as viscid vehicles for "mixtures." They are the mucilages, respectively, of *acacia*, *sassafras pith*, *tragacanth*, and *elm*. Being physiologically inert, these mucilages have no defined dose.

Next in order of simplicity among watery solutions are those that result from the treating of a crude drug with water until its virtues are dissolved out, and the subsequent rejection of the undissolved portions by straining. Such treatment with water may be effected in two distinct ways: in the one, the drug is actually *boiled* in the water, yielding a preparation which then is called DECO′CTUM, *Decoction*, while in the other it is allowed only to *steep*, the water at the time of addition being hot or cold, as the case may be. In the latter case the preparation is entitled INFU′SUM, *Infusion*, and in both these titles it must be noted that the Latin word is a *participle* (literally "a boiled thing," "an infused thing"), and not, as is the case in the English titles, a participial noun. In the present Pharmacopœia, general directions are given for the making of decoctions and infusions, so that the prescriber may order a decoction or infusion of any suitable drug he pleases. The official strength, unless otherwise directed, will be five per cent. of drug substance to a given amount of product. But besides giving such general directions, the Pharmacopœia also establishes by name a few special decoctions and infusions—special either by reason of variation from the above strength, or because of complex composition. Such special preparations are, severally, *decoction of cetraria*, *compound decoction of sarsaparilla*, *infusion of cinchona*, *of digitalis*, *of wild cherry*, and *compound infusion of senna*. Decoctions and infusions, unless gifted with powers of self-preservation, as is the case to a certain extent with infusion of cinchona, spoil readily, and should therefore, upon prescription, be freshly prepared. But since such doing is troublesome, it is to be feared that pharmacists often yield to the temptation to dispense a diluted *fluid extract*, when a decoction or an infusion is ordered, instead of the preparation prescribed. Decoctions and infusions are very much less used now than formerly, partly because of the above-described drawback of either delay or deceit in the dispensing, but, in main part, for the reason that the refinements of modern pharmacy have outrun the crudity of these bulky and bad-tasting "teas" of old, and have, in almost every instance where a decoction or infusion used to be prescribed, given us some other, and preferable, fluid representative of the drug in question.

The *average dose* of a decoction or infusion will range from two to four tablespoonfuls, but important reductions from this dosage will be required in the case of decoctions or infusions of potent drugs, as, for instance, in the case of the official *infusion of digitalis*, whereof a single dessertspoonful is the beginning dose.

Besides water, *alcohol*, *acetic acid*, and *glycerin* are used as medicinal solvents, because of special properties, affording especial advantages in certain cases, which these fluids severally possess. Of these three solvents alcohol is the most generally applicable, and proves, indeed, of great pharmaceutical value from the fact of its conjoining widely *solvent* with preservative powers. In very many cases, furthermore, compared, as a solvent, with water, alcohol possesses the double advantage of, on the one hand, dissolving more readily, and so extracting more perfectly, the active principle or principles from a drug, while, on the other, refusing to dissolve, or dissolving but sparingly, many of the other and undesirable constituents of the substance. Indeed, many medicinal vegetable principles which water will not touch at all are yet quite freely soluble in alcohol. Alcoholic solutions are separated into classes, as follows: the title TINCTU'RA, *Tincture*, applies to the analogues of "solutions" and "infusions" among watery preparations, *i.e.*, to direct solutions of salts or other solids in alcohol, and to the products of soakage of crude drugs in the fluid, whereby the alcohol dissolves out the virtues. By far the greater number of tinctures are of the latter type, in preparing which, in a very considerable number of cases, alcohol more or less diluted with water is used, for pharmaceutical reasons, instead of the undiluted article. Hence tinctures differ materially in alcoholic strength. Setting aside the two tinctures which are solutions of chemicals,[1] both of which are unique, the remainder—tinctures of vegetable and of animal drugs—form a distinct class of preparations, presenting, in general, the following characteristics: (1) They are *dark-colored*,

---

[1] Tincture of *ferric chloride* and of *iodine*.

whence the name *tincture*, signifying, literally, a something *tinged*. (2) They are *tenuous*, whence the fact that they yield small-sized *drops*, a point to be remembered in the directing of doses of tinctures by such measure. (3) They are *self-preservative*, although, from the volatility of the alcohol of their composition, they will readily, through imperfect corkage of the bottle, suffer the dangerous change of *over-concentration*. (4) As compared with the average decoctions and infusions, they are comparatively *strong*, and hence, in the case of tinctures of powerful drugs, the dose is relatively small, rarely exceeding a teaspoonful, and in very many instances being not more than a few drops. (5) They are, of course, *alcoholic*, and so, in the case of tinctures whose dose is of appreciable bulk, may be objectionable in conditions where alcohol is contra-indicated. (6) They are, as a rule, *less offensive to taste* than the aqueous preparations of the respective drugs, and are, in general, easy of administration. Tinctures are often prescribed as ingredients of composite mixtures, in which case, if mixed with aqueous preparations, regard must be had to possible differences of solvent property between alcohol and water. For if a substance in solution in a tincture be, as is the case with resins, insoluble in water, the same will precipitate on mixing such tincture with a watery preparation. Such precipitation, it is true, may not injure the medicinal activity of the substance so affected, but yet may be undesirable all the same. Seventy-one tinctures are official in the U. S. Pharmacopœia. A special class of tinctures, established, first, by the Pharmacopœia of 1880, is that entitled *Tinctures of Fresh Herbs*—TINCTURÆ HERBARUM RECENTIUM. In this instance, however, the Pharmacopœia does not direct preparations of individual drugs, but gives out only a general formula, according to which the pharmacist is to proceed upon receipt of any prescription for the tincture of a fresh herb. The physician is thus given the opportunity of ordering this kind of preparation in the case of any drug he may please, provided only such drug is one obtainable *fresh* in his locality. The pharmacopœial formula for these preparations directs the taking of one part of drug to two of alcohol. This class of tinctures is designed to meet the requirements of cases where the active principle of a plant suffers in the drying of the drug, either by reason of volatility or of inherent proneness to chemical change. Tinctures of fresh herbs have, however, the serious disadvantage of *uncertainty of strength*—a circumstance resulting from the fact that different samples of fresh herbs vary in the amount of water contained in their respective juices. Hence, of different lots of the same herb, equal weights may contain very unequal proportions of water, and, consequently, correspondingly unequal amounts of aqueously dissolved active principles.

SPI′RITUS, *Spirit*, vernacularly also *essence*, is a title applied to the alcoholic exact analogue of the "water," *i.e.*, to a preparation formerly commonly made by distilling alcohol from a drug holding a volatile principle, but now, as in the case of the "waters," most generally derived by direct solution of the previously isolated principle. The *spirits* of the Pharmacopœia embrace alcoholic solutions of *volatile oils* and of *camphor*, all made by direct solution; of certain *ethereal* bodies—ether, chloroform, nitrous ether, etc.; of *ammonia*, of *glonoin* (nitroglycerin), of *phosphorus*, and of the two distilled liquors, *brandy* and *whiskey*. Of these several groups, the spirits of the aromatic oils form a distinct class, naturally comparable with the "waters" prepared from the same substances. In such comparison there is an agreement—both sets of solutions are of peculiarly aromatic or fragrant oils; and a difference, the "waters" being weak, while the spirits are strong, for the reason that alcohol is a free instead of sparing solvent of volatile oils. The aromatic spirits, therefore, afford medicines as well as mere flavoring agents. Such spirits are the spirits, severally, of *bitter almond*, *anise*, *orange* (simple and compound), *cinnamon*, *gaultheria*, *juniper* (simple and compound), *lavender*, *lemon*, *peppermint*, *spearmint*, *myrcia* ("bay rum"), and *nutmeg*. Spirits, like tinctures, are valuable because of their concentration and keeping qualities, and, medicinally, are of advantage or of disadvantage, according to circumstances, from the alcohol of their composition.

VI'NUM, *Wine*, is the title where wine is used instead of diluted alcohol as a solvent. The wine in such cases is a natural weak "white wine," which, with the addition of a small amount of alcohol or of tincture of sweet orange peel, is then applied in the same way as alcohol in tinctures, viz., to dissolve chemicals, or to extract the virtues of vegetable drugs. Medicated wines are thus really a variety of tincture, and are but a poor variety at that, being less certain in strength and more liable to spoil than tinctures proper. There are but eight wines official in the Pharmacopœia, and of them perhaps the wine of *colchicum root* is the only one to be recommended as the most advantageous preparation of its drug.

Acetic acid, like alcohol, dissolves some things which water will not, and its solutions keep fairly well, though not so well as alcoholic ones. The acid, diluted, is used in two cases to extract the virtues of vegetable drugs, but in each case we have other preparations equal, at least, in value. The title of the product is ACE'TUM, *Vinegar*, and the strength a uniform one—virtues of ten per cent. of crude drug in a given quantity of preparation.

Glycerin is a unique menstruum, combining, like alcohol, extensive solvent powers with keeping properties, but, unlike that fluid, being, physically, viscid and non-volatile, and, physiologically, of mild taste and bland quality. Six glycerin-solutions are official, viz., of *carbolic acid*, *tannic acid*, *starch*, *boroglycerin*, *hydrastis*, and *yolk of egg*. The title of a glycerin-solution is GLYCERI'TUM, *Glycerite*.

Lastly among styles of fluid preparations for varied use comes the invaluable EXTRA'CTUM FLU'IDUM, *Fluid Extract*, which is an alcoholic extract concentrated by evaporation, and, unless self-preserving, fortified against change by some appropriate means, such as by the addition of glycerin. A unique peculiarity of fluid extracts is the strength of the preparation, which is uniform, and so ordered that the fluid extract shall exactly represent in a given *measure* the virtues of the corresponding *weight* of crude drug from which it is made—meaning by "corresponding" the correspondence of the metric system, *i.e.*, the weight of the same measure of *water*. In other words, specifically, a *cubic centimeter* of fluid extract is the medicinal equivalent of a *gramme* of drug. In our own more familiar wine-measure and apothecaries' weight the correspondence is not exact, simply because a fluidrachm is not the exact measure of a drachm-weight of water; but it is yet so near that, with the ever-existing latitude of dosage, it is perfectly legitimate, in prescribing, to reckon that a *minim* of fluid extract will equal a *grain* of drug; a fluidrachm a drachm, and a fluidounce an ounce. Fluid extracts are, then, fluid preparations as free as may be from inert or obnoxious constituents of the crude drug; keeping well, concentrated, and, in strength, bearing a uniform and simple relation to the strength of the original drug-substance. Because of these obvious advantages each revision of the Pharmacopœia has added largely to the list of fluid extracts, until now no fewer than eighty-eight are official. Fluid extracts are available, medicinally, for internal or external use, and, pharmaceutically, to serve as bases for other preparations. Given internally, the *dose* is always comparatively small, because of the high concentration of the preparation—with powerful drugs a single drop often being full allowance.

Passing now to the styles or literal *forms* of medicines which are designed for special applications, we find forms especially devised for giving by the *mouth*, others for use by the *rectum*, and others for application to the *skin*. For giving by the mouth we have, first, a form of powder, devised for the purpose of securing extreme fineness of pulverization and of yielding a powder of convenient bulk of dose in cases where the simple powder of the drug would prove inconveniently small. For this combination of purposes the scheme is to triturate a medicament thoroughly with a proper quantity of *sugar of milk*, a substance which combines the qualities of hardness of its particles, solubility in water, and agreeability of taste. For such a *dilute* powder the Pharmacopœia authorizes, under the title TRITURA'TIO, *Trituration*, the rubbing of one part of a powdered drug with nine of sugar of milk. The direction here is simply a general one, so that the prescriber has the privilege of ordering any powder he pleases to be made into a "trituration." The trituration is a convenient form for the giving of powdered drugs whose dose is small, and, because of the fineness of pulverization obtained, is the form of solid medicine that most nearly approaches a solution in speed of absorption. From the large proportionate bulk of sugar of milk in triturations, these preparations are not likely to taste unduly bad, and they are, as a rule, administered dry, upon the tongue.

In the reverse set of cases, where, instead of a dilute, a concentrated powder is desired, we have the indication met, so far as they go, by those *extracts* which are capable of pulverization.

In the case of children or of squeamish adults, it is always desirable to conceal offensive taste in a medicine. One way of so doing, with powders, is to incorporate the material in a sugary mass having the agreeable qualities of soft confectionery—a method which meets the indication, it is true, but does so at the disadvantage of throwing upon an invalid stomach those obnoxious matters, *sugars*, in considerable quantity. However, we find in the Pharmacopœia two preparations of this character: technical name CONFE′CTIO, *Confection.* One is a *confection of rose*, being simply an agreeably flavored confection-mass to serve as the basis for confections extemporaneously prescribed, and the other is a laxative *confection of senna.* Confections are given to eat, like candy, the charge of medicament being small.

The well-known *pill*, the form next to be discussed, finds place in the Pharmacopœia under title PI′LULÆ, *Pills*, and MA′SSA, *Mass*—that is, pill-material before subdivision. The terms *granule* and *parvule*, often applied to very small pills, are vernacular only. Here is, in the literal sense of the word, a *form* of medicine availed of by the physician for extemporaneous prescription, as well as by the Pharmacopœia for a set formulary of preparations. The pill-form presents many peculiar features for consideration, some of advantage and some the reverse. Of advantage are permanence, portability, exactitude and convenience of dosage, and concealment of bad taste; while of disadvantage are comparative slowness and uncertainty of absorption of the contained medicament, difficulty or impossibility of administration to many persons, including obviously the entire class of *little* children, who constitute so very large a proportion of our patients. Convenient, therefore, as pills are, they must not be prescribed with stupid indiscrimination. Almost any solid medicine not deliquescent, in powder or as extract, and also many fluids, provided, of course, the dose be small, may be ordered to be dispensed in pill-form. Pills when freshly made are dusted with some dry powder to prevent them from cohering, or are coated with some material with the view of concealing taste. For the latter purpose a simple and handy process, applicable to small batches of pills, is to shake the pills, freshly made and still sticky, in a box with gold or silver foil. By this means the pills become loosely coated with bits of the broken foil—a covering which fairly enough conceals the taste and yet readily gives way, after swallowing, so as to interpose no obstacle to the solution of the pill in the stomach. Other substances so largely used for coating pills—sugar, gums, gelatin—require special manipulation and even apparatus for their application, so as to be practically available only for the manufacture of a considerable number of pills. A batch of a dozen or so extemporaneously prescribed pills can, therefore, hardly, with profit to the dispenser, be coated with any of these materials. These coatings have furthermore the feature that the time required for their dissolution in the stomach is just so much time lost for the operation of the pill, and if such time be considerable the pill may slip along the alimentary canal even into the lower bowel, before its own solution begins, and the full absorption of its medicament thus be seriously compromised. Hence, without touching the question of the relative ease of dissolution of these various coatings, it is a good general rule that when speed and certainty of operation are desirable, coated pills (coated otherwise than by metallic foil) had better not be allowed. Besides coatings for the concealment of taste, some pills, because of the nature of their ingredients, require an air-tight casing, as, for instance, pills containing phosphorus or ferrous iodide, substances that easily oxidize on exposure. Balsam of tolu is used for such coating, and the pills thus prepared are open to the same possible objection as just urged against other coated pills.

As regards the administration of pills, the majority of persons old enough to take a pill at all can readily swallow the little sphere if put far back upon the tongue and helped along by a gulp of water. But if there be reflex objection on the part of the surprised pharynx, encase the pill in some slippery mass—chewed bread-pulp, apple-scrapings, or a bit of preserve; or—a method found to succeed when all others fail—take a dark-skinned grape, in which the pulp slips easily from the skin, dig out the seeds, put the pill in their place, and then give the grape to be eaten in the way so commonly done, *i.e.*, by popping the pulp into the mouth and swallowing at once without chewing. But in spite of every device, some persons, even adults, can never swallow a pill—the mere knowledge of the pill's presence in a pulpy mass sufficing to determine the involuntary rejection of the same.

The pill being such a favorite form of medicine, we have quite a number of *pharmacopœial pills*, some simple, some compound. In three instances a pill-mass only is ordered, leaving it to the prescriber to direct the weight of the individual pills, but in all other cases the Pharmacopœia establishes the weight of the pill as well as the composition of the mass. Officially a pill-mass is entitled MA′SSA, *Mass;* and pills, PI′LULÆ, *Pills.* The three pharmacopœial "masses" are those, severally, of *copaiba*, *ferrous carbonate*, and *mercury* ("blue mass").

The pill is one of the forms very commonly selected for the extemporaneous prescription of appropriate drugs. In so prescribing, regard must be had to the points already made concerning what medicines and *what patients* may properly enter into pill-relation. But also now presents itself the subject of *excipients* for pills. The very condition of the pill necessitates a certain quality of stickiness, and this, it needs not to say, belongs to but few of the things to which we wish to give the pill-form. Some excipient, therefore, will be required in the majority of instances of pill-prescriptions, and this excipient will vary according to the physical and chemical qualities of the basis. Considerable knowledge of pharmacy is thus involved in the proper fitting of excipient to basis, and because of this, and because, nowadays, the physician is not expected to be also a pharmacist, it is the practice with many, in prescribing extemporaneous pills, simply to order that so much of a given medicine shall be made into so many pills, leaving it entirely to the compounder to take what and how much excipient pharmacy knows to be best for the case. But since many physicians, on the other hand, prefer, in prescribing pills, themselves to direct the excipient, it is proper to point out here the general principles governing the selection. *Sticky vegetable extracts* require no excipient, and, furthermore, those of feeble medicinal power make, themselves, capital excipients for *heavy powders*; for example, extract of gentian as the excipient for reduced iron. If a little too firm, a few drops of water will effect the necessary softening. *Soft gum-resins* need no excipient, or at most a few drops of alcohol to reduce hardness. *Semifluid* or *fluid* substances require some indifferent dry powder, such as powdered gums or starches. Bread-crumb, wheat starch, or gum-arabic are the substances most commonly used for the purpose. *Powders*, if *heavy*, such as metallic compounds in powder, may be mixed with a soft vegetable extract, or (unless chemically incompatible with tannic acid) with confection of rose. *Light powders*, such as simple vegetable powders, make up best by moistening with some viscid fluid, such as syrup, honey, or glycerin. The latter substance, because peculiarly non-drying, is specially advantageous if it be desirable that the pills be kept soft for some time. On the other hand, *mucilage* is objectionable as an excipient, because pills made by the use of this material speedily become hard—too hard for certain solution in the stomach. *Resinous and fatty bodies* do well by admixture with *soap*. In any case, in prescribing the excipient it is enough to order the *selection* only, leaving the question of *amount* to the compounder. A special style of pill, possible to make with appropriate apparatus, is the so-called *compressed* pill—a pill alleged to contain little or no excipient, but to consist simply of the medicament, in dry powder, whose particles are made to cohere, so as to retain the pill-form, by the application of powerful compression. Compressed pills are liable to the fault of being too hard—so hard, that is, as possibly partially, or even wholly, to escape dissolution in the alimentary canal.

In the extemporaneous prescription of pills, a necessary point to bear in mind is that of the *bulk* of the pill. In general, pills should be of small size only; and hence, as a rule, there should not be ordered, to constitute a single pill of light bodies, such as vegetable powders, more than *five grains* (thirty centigrammes), or, of heavy substances, in excess of *six or seven grains* (from forty to forty-five centigrammes). And better is it, indeed, not to exceed the one-half of these weights, respectively.

TROCHI'SCUS, *Troche*, is the technical name of the well-known *lozenge*, applied as a form of medicine. Troches are designed to be held in the mouth and sucked until dissolved, and are resorted to, mainly, as affording a convenient way of continuously medicating the oral or pharyngeal cavity in surface affections of those parts. Under the circumstances slowness of solution—contrary to what obtains in the case of pills—is here an advantage, and hence we find *tragacanth* as the gummy basis of a majority of the official troches. Troches are pleasant to take, and, besides their more natural purpose, as above, are often used as the form for medicines aimed to relieve cough—many of the official and numberless of the proprietary troches being compounded for this special application.

Partaking of certain of the properties of the pill on the one hand, and of the troche on the other, is a form of medicine which, although not official, is nowadays very considerably used. Such form is what is commonly styled *tablet*. The "tablet" is a thick disc of small size—about the diameter of the average pill—and commonly consists of some soluble medicine made into tablet-form by admixture with some soluble excipient. The tablet is constituted of soluble ingredients, for the reason that the special purpose of the preparation is to afford a means of having ready to hand accurately determined doses of a medicine to be administered *by solution*. The majority of the tablets of the market dissolve readily in water, cold or warm, and the convenience of the preparation is obvious. Besides their legitimate purpose of thus serving as a means for the extemporaneous preparation of a dose of a medicine to be given in solution, tablets may also be used as substitutes for pills, since the little discs are readily enough swallowed whole by any one who can take a pill.

Lastly, among preparations specially intended for use by the mouth, are certain *fluid* forms, devised to secure agreeability of taste by the presence of considerable sugar. Where the peculiar feature is simply such addition of sugar to other ingredients, the preparation is termed SYRU'PUS, *Syrup*. A considerable number of syrups are official, the group embracing syrupy solutions of salts and other inorganic substances, as well as of vegetable drugs—in certain of the former kind the sugar being of more importance as a preservative than as a flavoring agent. These same syrups of inorganic matters are too incongruous to present any general class-characteristics, but the syrups derived from vegetable drugs form a fairly distinct group. Such syrups are variously made by the addition of sugar or of syrup to expressed juices, solutions, "waters," infusions, decoctions, "vinegars," tinctures, and fluid extracts. They are of course sweetish to taste, and so are pleasanter than the average fluid medicines, but, medicinally, they rate comparatively low in strength, and are not to be resorted to where concentration of dosage is desired. Rarely is the dose less than a teaspoonful[1] and often it is a tablespoonful or more. A number of official syrups, furthermore, have no, or practically no, medicinal power whatever, and are offered simply as agreeably flavored matters to constitute part of the fluid vehicle in extemporaneous prescriptions. When so used, these syrups should, as a rule, not form more than one-half the volume of the mixture, else the potion will be too sweet unless diluted at

[1] Notable exception, *compound syrup of squill.*

the taking. Among such syrups is the preparation called by the single title *Syrupus*, "Syrup," a simple aqueous solution of cane-sugar, of specific gravity 1.317. The syrups available as agreeable flavoring agents are "syrup," and the syrups, severally, of *citric acid*, *almond*, *orange*, *orange-flowers*, *wild cherry*, *rose*, *raspberry*, *tolu*, and *ginger*. The syrups, respectively, of *acacia* and *althæa* are mucilaginous as well as syrupy.

Concerning all syrups a final point needs to be made that the preparations are more or less prone to change. Often in the domestic medicine-chest a long-kept bottle of syrup, as of ipecac, will be found with the cork blown out and the fluid contents turbid and frothy. These conditions are the results of fermentation, one of the commoner of the modes of decomposition to which syrups are liable. Hence in prescribing syrups, or mixtures into which a syrup largely enters, the rule is to order no more of the preparation than is likely to be used for the case in hand.

Practically a syrup is the single preparation of the Pharmacopœia entitled MEL, *Honey*, viz., *Mel Rosæ*, "Honey of Rose." The preparation consists simply of honey impregnated with the flavor and mild astringency of red rose.

Of late a favorite mode, with manufacturing pharmacists, of catering to the popular passion for toothsome medicines, has been to offer a fluid composition containing a little of some drug-principle and a good deal, severally, of alcohol, sugar, and aromatic flavoring. Such a preparation the manufacturers have entitled ELI'XIR, *Elixir*. Recognizing the hard fact of the popularity of these elixirs, the Pharmacopœia has thought it no more than fair to offer to the legitimate prescriber a ready means of competing with the wholesale manufacturer in this field. We find, therefore, official a so-called *Aromatic Elixir*—simply dilute alcohol, sweetened and flavored with the oils, respectively, of orange peel, lemon, coriander, and anise—which elixir may be used as a vehicle for the making of medicated elixirs, either by dissolving substances directly therein, or by charging it with the proper quantity of a tincture or fluid extract. But in prescribing this elixir-basis, it must be borne in mind that the same is nearly *twenty-five per cent. alcohol*—is stronger, that is, in alcohol, than the strongest sherry wine. Medicate it weakly, then, as is the way with elixirs; order it in tablespoonful doses, as is the necessity in the case of weak mixtures; let it be taken regularly for a month or two, as is the rule with "tonic" medicines, and then be not surprised if the whiskey-bottle succeeds the elixir-vial on the shelf of the patient's private closet. There is also official one really medicinal elixir, viz., an *Elixir of Phosphorus.*

To medicate the *rectum*, the *vagina*, or the *urethra*, we have the simple device of incorporating the medicament with a material which, while liquefying readily at the temperature of the body, is yet firm enough to admit of being passed bodily, in form of a solid plug, into one of these canals. All such medicated plugs are, generically, entitled *suppositories*, and under the simple title SUPPOSITO'RIA, *Suppositories*, the Pharmacopœia establishes a general method for the making of suppositories of any drug which the physician may prescribe, and for any one of the three applications just mentioned. The excipient directed is cacao butter ("oil of theobroma"), a substance that perfectly meets the requirements. According to the Pharmacopœia, "unless otherwise specified, Suppositories should have the following weights and shapes, corresponding to their several uses: *Rectal Suppositories* should be cone-shaped, and of a weight of about *one* (1) *gramme. Urethral Suppositories* should be pencil-shaped, and of a weight of about *one* (1) *gramme. Vaginal Suppositories* should be globular, and of a weight of about *three* (3) *grammes.*" Under these directions, all that the prescriber has to do is to direct that the pharmacist take a certain amount of designated drug and make it into a certain number of suppositories of specified kind. A single specially medicated kind of suppository is official, namely, *Suppositories of Glycerin.* Rectal suppositories are the kind most commonly used, and, concerning their application, the only points are that the medication will be more thorough the cleaner the cavity that receives it, and that, in insertion, the plug must be pushed up beyond the sphincter.

To medicate the *skin*—or the system at large through the avenue of the skin—we have a number of special pharmaceutical forms. It is here often most suitable that the medicine be incorporated with a *fatty* substance, for the reason that greasy dressings protect from the air, prevent drying, and, more readily than water-moist matters, permeate cracks, crannies, or even the unbroken tissue of the skin. Medicated fatty mixtures give us three classes of preparations, as follows:

UNGUE'NTUM, *Ointment*, is the title when the substance is of soft, lard-like consistence, suitable, when so needed, for inunction. The most commonly-used bases for ointments are, severally, *lard*, or lard with a small admixture of wax slightly to increase its firmness, and the well-known singular substance *vaseline*, so-called.

Pure lard readily turns rancid, but this tendency may be nullified by impregnation of the fat with benzoin. Vaseline is peculiar—and hence peculiarly valuable—in its quality of unalterability; regardless of exposure it neither spoils nor dries, and it is singularly indifferent to the action of even powerful chemical agents. All of these bases are official under the several titles, *A'deps*, "Lard"; *A'deps Benzoina'tus*, "Benzoinated Lard" (lard preserved by impregnation with benzoin); *Ungue'ntum*, "Ointment," a mixture of lard and wax, four parts of the former to one of the latter; and, designating vaseline of different degrees of consistence, *Petrola'tum Li'quidum*, "Liquid Petrolatum"; *Petrola'tum Mol'le*, "Soft Petrolatum"; and *Petrola'tum Spis'sum*, "Hard Petrolatum." According to the Pharmacopœia, "when Petrolatum is prescribed or ordered without further specification, Soft Petrolatum (*Petrolatum Molle*) is to be dispensed." Other soft, fatty substances serviceable as bases for ointments are *Se'vum*, "Suet," and *A'deps La'næ Hydro'sus*, "Hydrous Wool Fat." The latter substance consists of "the purified fat of the wool of sheep, mixed with not more than 30 per cent. of water." It is a soft, fatty material, capable of being mixed with double its own weight of water without losing its unctuous quality. The Pharmacopœia also offers a number of medicated ointments, the basis, in the majority of cases, being benzoinated lard.

CERA'TUM, *Cerate,* from Latin *cera,* wax, signifies a preparation analogous to an ointment, but firmer—so firm as not to melt at the temperature of a warm skin, and intended for use as a more or less permanent dressing, applied by being spread upon muslin or other backing. The basis is a fat, either lard or oil, and the peculiar consistence is obtained by admixture of either wax, spermaceti, or resin. A few cerates are official: among which one, entitled simply *Ceratum,* "Cerate," is a mere mixture of lard and wax (seven to three), convenient as a basis for extemporaneously medicated cerates.

The third kind of preparation with fatty basis is peculiar. Fats can be decomposed by chemical means into acid and basic radicals, one of the former of which is the body *oleic acid,* a thin, oily fluid, having the physical attributes of the fixed oils. This body, being a true acid, can unite with bases to form salts—*oleates,* as they are then called, in accordance with chemical nomenclature; and it so happens that certain of such salts are readily soluble in an excess of the acid. Such soluble oleates are notably the oleates of metallic bases and of alkaloids, and so, since these basic bodies are common and important medicines, pharmacists have devised as a kind of preparation of them a solution of the respective oleates in oleic acid. Such a preparation is officially entitled OLEA'TUM, *Oleate*—a title well enough in the Latin, but, in the English, likely to breed confusion, because thus the pharmaceutical name of a specific solution of a salt is made identical with the simple chemical name of the salt itself. These so-called "oleates" have been devised because of the peculiar property which oleic acid possesses, and with which it endows its salts, of *permeating tissue* with extraordinary readiness. "Oleates," therefore, are more reliable than ointments for the medication, by inunction, of such subcutaneous tissues as are deep-seated or are unusually dense. Theoretically, also, these preparations surpass ointments for the purpose of medicating the *blood*, and so the system generally, by the method of inunction. Practically, however, they do not always prove satisfactory in this application, since some clinicians have reported failure to mercurialize by inunction with oleate of mercury in cases where subsequent success followed the resort to mercurial ointment.

An objection to "oleates" is a tendency these bodies have to irritate, because of the irritant nature of the free oleic acid of their composition.

These preparations, therefore, should be but lightly applied, and, in the case of sensitive skins, should be dosed with one per cent. of morphine—the alkaloid, not any salt of the same—before application. Such charge of morphine will dissolve by combining with some of the free oleic acid of the preparation to the formation of a soluble oleate of the alkaloid.

The Pharmacopœia recognizes three "oleates," viz., those, respectively, of *mercury*, of *zinc*, and of *veratrine*, whereof the first is of twenty, the second of five, and the third of two per cent. strength of dissolved base. Other "oleates," however, are offered by manufacturing pharmacists. According to their strength and to the nature of the base in an "oleate," these preparations vary in consistence from that of a thin oil to that of a soft fat, such as lard.

Besides "oleates" such as the foregoing, where the preparation consists of an oleate proper in solution in an excess of oleic acid, certain oleates—using the word now in its proper chemical sense—are articles of manufacture and are of use for local medication of the skin. The oleate of *lead* is an example in point. These oleates, contradistinguished from the pharmacopœial preparations styled by the name, are dry, pulverulent bodies, of a smooth, soapy feel.

EMPLA'STRUM, *Plaster*, as the title is used in the Pharmacopœia, means a stuff, in mass, proper for spreading upon some backing to make a "plaster," as ordinarily understood. For their purpose plasters need to be sticky and firm, qualities found pre-eminently in a peculiar material that results from the boiling together of litharge, olive oil, and water. This material, known as "lead plaster," is thus a favorite basis for these preparations, other bases being various admixtures of resins, gums, waxes, and fats. Plasters are hard at ordinary temperatures, and require softening by heat for spreading. For use the material is spread in a thin layer upon sheepskin or other leather, or upon linen or muslin cloth, and, for application, may or may not require to be somewhat softened by gentle heat. When once upon the skin, plasters stick tight, resist water, except in the case of "court-plaster," and, for removal, must be stripped off by force. If hard to start, a corner may be loosened by moistening with oil of turpentine. If a part be hairy, the same should be shaved before application of the plaster. For specific medication plasters are at best feeble and many of them are inert. What action, therefore, plasters have is mainly the physical one of affording a certain amount of mechanical support, while at the same time air is excluded from the part and a gentle local irritation maintained. Some plasters, indeed, such as *adhesive* and *court* plasters, are for no other purpose than to afford protection and coaptation of parts.

Plasters are prescribed by dimension, in inches, and not by weight, and are dispensed by the pharmacist ready spread upon proper backing.

LINIME'NTUM, *Liniment*, is a name applied generically to any more or less distinctly *fluid* preparation (except "oleates") intended for rubbing upon the skin. The pharmacopœial liniments are so very incongruous as to present no class-features for present discussion.

Lastly among pharmacopœial preparations comes CHA'RTA, *Paper*, meaning, naturally, a medicated paper. Two such "papers" are official. Of these, one—namely, the paper of *mustard*—is intended for local application to the skin, and consists of paper coated on one surface with a preparation of the drug. The other, paper of *nitrate of potassium*, consists of a bibulous paper impregnated throughout its texture with nitre, and is intended to be burnt for the sake of evolving medicated fumes for inhalation.

Finally, to close the subject of forms of medicines, we may note here the following technical terms, which, although not occurring in pharmacopœial nomenclature, are yet in common use in the present relation; *bougia*, "bougie," a urethral suppository; *ca'psula*, "capsule," already commented on; *catapla'sma*, "poultice"; *cha'rtula*, "little paper," meaning, in prescription-writing, the separate paper package into which each dose of a powder is to be put up; *colly'rium*, "eye-drops"; *discus*, "disc," generally a thin disc of gelatin, medicated, for application to the eye or for solution for hypodermatic injection; *e'nema*, "enema," "clyster," a rectal injection; *gargari'sma*, "gargle": *hau'stus*, "draught," a considerable potion to be swallowed at a dose; *inje'ctio*, "injection"; *lo'tio*, "lotion," a wash; *pedilu'vium*, a foot-bath; *su'ccus*, "juice," formerly official, meaning the expressed fresh juice of a vegetable drug preserved by the addition of a little alcohol.

CHAPTER IV.

THE DETERMINING OF QUANTITIES OF MEDICINES.

THE subject of the *determining of quantities* of medicines concerns the physician, first, in the matter of the *compounding of extemporaneous prescriptions*, and, secondly, in that of the *dispensing of doses*. For one or the other of these purposes both of the two several methods, *weighing* and *measuring*, are employed, so that the technology of both of these methods requires consideration.

Of the determination of quantities of medicines by weight.—As a process, weighing has the advantage over measuring of being intrinsically capable of greater accuracy of achievement, but yet is subject to the disadvantage that it requires special apparatus and skill for its accomplishment. Hence, since patients cannot be expected to own and operate balances, weighing is entirely ruled out from application to the important purpose of the *domestic dispensing of doses* in the case of medicines ordered in bulk, whether fluid or solid. The resort to the method of weighing for the determining of quantities of medicine is, therefore, practically limited to the operations of the pharmacist in the compounding of prescriptions. For this purpose the greater accuracy of the method commends it; and, furthermore, for one who, like the pharmacist, owns and is an adept at the balance, determinations by weighing are quicker and more convenient of accomplishment than by measuring. This fact is obvious in the case of dealings with solids, but even in the case of fluid medicines the balance outvies the measuring-glass in convenience, while far surpassing it in accuracy. For it is only necessary to counterpoise, upon a pan of the balance, the bottle in which a prescribed mixture is to be dispensed (an operation done in a few seconds by an experienced hand), when, by the consecutive addition of the proper weights, one after another of the fluid constituents of the prescription can be poured directly from the stock-bottle into the dispensing vial, the pouring, in each instance, being arrested at the drop that tips the balance. Such operation is both rapid and accurate, and entirely does away with that nuisance that attends the mensuration of fluids—the washing-up of dirty graduate-glasses.

For these various reasons many pharmacists

prefer to compound by weight not only in the case of solids, but also in that of fluids, and such, therefore, of course, desire that the physician prescribe his quantities in terms of weight. Now, this the prescriber is willing enough to do in the case of *solid* drugs, since *doses*, in the case of solid medicines, are to be dispensed, and are, therefore, learned, by weight. But in the matter of *fluid* constituents of a prescription there arises an issue between prescriber and compounder, for the following reason: In the greater number of instances where a fluid medicine is prescribed, the final product is to be a bottle of fluid, for internal giving, by doses to be doled out at the bedside. Now, such doses, it hardly needs to be said, are to be *measured* from such bottleful, and not *weighed*. In the prescribing, therefore, the physician has in mind simply an aggregate of *volumes*, and must apportion his ingredients by measure of volumes only. In such case, then, if the prescription is to be filled by means of the balance instead of the "graduate," it comes to the prescriber's ordering *volumes* by *weight*; which means that, having thought out his volumes, he is to find out what the same will weigh, and then write for the quantities by the figures thus calculated. Now, since no two fluids weigh just the same, measure for measure, this translation from one system of determining to another, if to be done with severe accuracy, requires a knowledge of endless diversities of specific gravity, and entails tedious computations for every determining of a quantity. And such seeming trouble and tedium it is that constitutes the theoretical objection of the prescriber to the plan in question.

Practically, therefore, in the matter of prescriptions, while weighing is universally applied in dealing with solids, it is otherwise, in this country, in the case of fluids. Yet, as Americans are now daily learning, by foregoing an extreme of accuracy the difficulty from diversity of specific gravities shrinks to trifling proportions, and then, *if only there be at hand a perfectly correlated system of weights and measures*, it is just as easy to set down quantities of volume in terms of weight as in terms of measure. Now, such correlation obtains in the so-called *metric* system, original in France, but now standard also in the majority of the enlightened nations of the world. Because of this correlation, then, it has naturally come about, being of advantage to the compounder, that, where the metric system is used, fluids as well as solids, among medicines, are prescribed and "put up" by weight. This same metric system, being legalized in the United States, is at the present day considerably used by the physicians of the country, both native as well as foreign, so that the practical procedure of prescribing volumes by weight demands consideration.

Now, the sole difficulty of the procedure arises, as we have seen, from the diversity of specific gravity among fluids, and presents itself thus: The correlation of metric weights and measures of capacity is through the medium *water;* the weight of a standard volume of this fluid (one cubic centimeter) being taken as the basis for the system of weights (one gramme). With *water*, therefore, there is absolutely no trouble, the same figure expressing quantity in terms of weight and of volume, both. This being so, let us look at our fluid medicines from the point of view of their *specific gravities as compared with that of water.* A very large number of such medicines are themselves of aqueous basis, or, from the proportion of their ingredients, show a gravity but little different from that of water—a difference so slight that the error introduced by disregarding it altogether is within the *error of dosage* (*i.e.*, the amount more or less than a given quantity whose physiological effect is inappreciable). Hence the practice—and a perfectly legitimate one—to ignore altogether the individuality of gravity among these medicines, and to treat them as if they weighed the same as water. And so, behold!

in practical working, just as in the case of water itself, the bugbear of specific gravity is shorn of its terrors in the cases, severally, of *infusions, decoctions, waters,* and of most *solutions, fluid extracts,* and *tinctures*—in short, of the majority of fluid medicines whose prescription requires an original computing by volume. There remains, though, a considerable number of fluid things on the list of the materia medica whose specific gravity differs so considerably from that of water that, in ordering weights with a view to getting volumes, the gravity *must* be taken into account. These things, however, divide, for the present consideration, into three categories. First, things which, while presenting a striking individuality of specific gravity, yet, from the nature of their medical relations, *rarely enter into the composition of fluid mixtures for internal giving,* and hence, in prescription, do not need to be thought of by volume. Thus we wish for a small bottle of strong nitric acid for surgical purposes: let us order half an ounce by weight, and though, because of the comparatively high specific gravity of the acid, such weighed quantity will be decidedly less than half a fluidounce by measure, yet what matters it for the purpose of the prescription! Similarly, a half-pound can of ether, ordered for anæsthetic use, may measure what it please—its volume is of no practical moment; and chloroform, though half as heavy again as water, is yet rarely so prescribed in combination as to require that cognizance be taken of the volume of a given weight.

The second category embraces medicines which, though combined in prescriptions, yet necessarily occur therein in such small proportions, as compared with the bulk of the bottleful, that a trifling error of quantity, when divided, as it comes to be, by the number of doses, falls again within the "error of dosage," and may therefore be disregarded. Into such category we may put, as one group, the *volatile oils*, the *spirits*, *tinctures made from alcohol* as distinguished from those where diluted alcohol is employed, and certain of the *fluid extracts*. Measure for measure, these fluids are one-tenth lighter in weight than water, and hence to get one cubic centimeter in volume we should, in accuracy, write for ninety centigrammes only, in weight. On the other hand, a few *fluid extracts* are appreciably *heavier* than water, but all the classes of preparations named are prescribed, in combination, in such small relative proportion that the several corrections for specific gravity may safely be, as they commonly are, quietly omitted in practice.

There is left, then, of the bone of contention

only the small piece occupied by the third category —that, namely, that consists of things distinctly different in specific gravity from water, and yet such, in their nature, that if prescribed in combination they will occupy a goodly or even the greater part of the bottleful. With these fluids, then, there is no escape; we *must* allow for their specific gravity when ordering volumes by weight in composite prescriptions. But behold! the list is but the trifling one here displayed, and the corrections, as the table shows, are the simplest things possible:

TABLE OF CORRECTIONS FOR SPECIFIC GRAVITY.

(As practically required for application in the prescribing of volumes by weight, in composite prescriptions.)

To get 1 C.c. of—	Order, in grammes.
Oils	0.90, or $\frac{1}{10}$ less
Glycerin	1.25, or $\frac{1}{4}$ more.
Syrups and honey	1.33, or $\frac{1}{3}$ more.

So, then, the whole of this theoretical mountain of difficulty of prescribing fluids by weight shrinks to the triple-peaked molehill that in ordering of oils you must write for one-tenth less, of glycerin one-quarter and of syrups one-hird more, than your estimated volume.

As thus applied, then, by the help of the peculiarities of the metric system the prescribing of fluids by weight is perfectly easy. The really objectionable feature of the system is that it introduces into prescriptions calculated upon a basis of volume an error commensurate with the specific-gravity value disregarded. But apart from the question whether this error is of practical importance or not, it is probably compensated for by the greater accuracy by which weights can be determined—the exactness of the compounding atoning, so to speak, for the inexactness of the prescribing.

Next we pass to the consideration of the different *scales of weight* used in dealings with medicines between physician and pharmacist. Three such only are likely to come under our cognizance. They are, first, the *apothecaries' weight*, consisting of the grain, the ounce, and the pound of troy weight, with the intercalation of two special denominations between the grain and the ounce; secondly, a *mixed* weight made standard by the British Pharmacopœia, consisting of the troy grain and the avoirdupois ounce and pound; and, thirdly, the weights of the *metric* system. Of these weights the apothecaries' is the one commonly used with us; the British pharmacopœial weight is peculiar to Britain, and the metric weight is standard in almost all enlightened nations except the United States and Great Britain, and even in such countries is already the system of science, and is daily making strides into favor also in the prescribing of medicines.

Following is an exhibit of apothecaries' weight, the table also giving the Latin names of the denominations and the symbols therefor used in prescription :

TABLE OF APOTHECARIES' WEIGHT.

Grain (*Gra'num*).	Sbruple (*Scru'pulus*).	Drachm (*Dra'chma*)	Ounce (*U'ncia*).	Pound (*Li'bra*).
Gr.	℈	ʒ	℥	lb.
20	=1			
60	3	=1		
480	24	8	=1	
5,760	288	96	12	=1

In this system the grain, the drachm, and the ounce are the denominations most commonly used. The actual weights are convenient for prescription-purposes, and the numbers expressing the respective relations of the denominations just named—namely, 1, 60, and 480—permit of easy calculatings. The objections to the system are the irregularity and non-decimal character of these same relations, which, despite the intrinsic advantages of the numbers 60 and 480, make the use of the scale, as compared with a decimal one, slow and cumbrous. There is also wanting any exact correlation between these weights and any standard measures of capacity.

The peculiar weight of the British Pharmacopœia is thus shown in tabular form :

TABLE OF BRITISH PHARMACOPŒIAL WEIGHT.

(Troy grain, avoirdupois ounce and pound.)

Grain (*Gra'num*).	Ounce (*U'ncia*).	Pound (*Li'bra*).
Gr.	oz.	lb.
437.5	=1	
7,000	16	=1

This system has the disadvantage that the ounce is an odd number of grains, but it possesses the two advantages, to offset, first, that it is the same system by which drugs are bought and sold commercially ; and, secondly, that it is, partially at least, correlated to a scale of measures of capacity, the ounce being the exact weight of a fluid-ounce (imperial measure) of water, and the pound weighing one-tenth of an imperial gallon. As compared with apothecaries' weight, it must be noted that while the British grain is the same, the *ounce* and *pound* of British weight, though identical in name, differ in value from their troy namesakes, the ounce being less and the pound more, as shown by the column of grain equivalents in the table. In prescribing in Britain, though the use of the apothecaries' (troy) ounce and of the pound is discouraged by the British Pharmacopœia, the use of the *scruple* and of the *drachm* is sanctioned by the same authority. Hence the symbol "ʒ" in British writing means the *apothecaries'* drachm of sixty grains, and *not* the *avoirdupois* drachm of twenty-seven grains and a fraction, the avoirdupois drachm not being recognized in the British Pharmacopœial weight.

The *metric* system has the enormous advantage that the denomination-ratios are identical and decimal, thus conforming, as in the case of our American currency, to decimal notation, and so reducing calculations to the extreme of ease and simplicity. In using the metric system we commonly speak only of *milligrammes*, *centigrammes*, and *grammes*, expressing other denominations in terms of these, just as in the case of our currency we, in ordinary speaking, ignore dimes and eagles, and numerate all amounts in dollars and cents only. In writing metric amounts, again as in

TABLE OF "METRIC" OR FRENCH DECIMAL WEIGHTS.

(More or less in use, universally, in the prescribing of medicines.)

Milli-gramme	Centi-gramme.	Deci-gramme.	Gramme (*Gram-ma'rium*).	Deca-gramme.	Hecto-gramme.	Kilo-gramme	Myria-gramme.
Gm. 0.001	Gm. 0.01	Gm. 0.10	Gm. 1.00	Gm. 10 00	Gm. 100.00	Gm. 1,000.00	Gm. 10,000,000
10	= 1						
100	10	= 1					
1,000	100	10	= 1				
10,000	1,000	100	10	= 1			
100,000	10,000	1,000	100	10	= 1		
1,000,000	100,000	10,000	1,000	100	10	= 1	
10,000,000	1,000,000	100,000	10,000	1,000	100	10	= 1

the case of dollars and cents, we use simply the ordinary arithmetical decimal notation, understanding the integer as grammes, the hundredths as centigrammes, and the thousandths as milligrammes.

An independent peculiarity and advantage of this system of weights is that the same is correlated to an analogous system of measures of capacity, the weight of a standard volume (cubic centimeter) of *distilled water*, at its temperature of greatest density, being the *gramme*—the unit of weight. This correlation helps enormously in that translation from estimate by volume to order by weight, which is involved, as already seen, in the plan of prescribing fluids by weight.

Since the metric and apothecaries' weights are both in use in American prescribing, we must know the relative values of the respective denominations of the two systems. Exactly, the gramme is 15.43234874 grains, but with the latitude always inherent in dosage, and therefore in prescribing, it is near enough for our purpose to take the equivalence, with small sums, at 15 grains, and with large at 15.5. For mutual translation between the systems the following table will be found convenient. It is purposely made short, in order that it may, as it easily can, be committed to memory. From this table, then, the equivalents of intermediate values can easily enough be calculated mentally. The *approximate* metric equivalents are accurate enough for prescription purposes, the exact values being added for information and not for use:

TABLE OF EQUIVALENTS.

Apothecaries' and Metric Weights.

APOTHECARIES'.	METRIC. (Approximate.)	(Exact.)
Gr. $\frac{1}{64}$ =	0.001 Gm.	[0.00101 Gm.]
Gr. $\frac{1}{32}$ =	0.002 Gm.	[0.00202 Gm.]
Gr. $\frac{1}{16}$ =	0.004 Gm.	[0.00405 Gm.]
Gr. $\frac{1}{12}$ =	0.005 Gm.	[0.00540 Gm.]
Gr. 1 =	0.06 Gm.	[0.06480 Gm]
Gr. 10 =	0.65 Gm.	[0.64799 Gm.]
Gr. 15 =	1.00 Gm.	[0.97198 Gm.]
℈ 1 =	1.30 Gm.	[1.296 Gm.]
ʒ 1 =	4.00 Gm.	[3.888 Gm.]
℥ 1 =	30.00 Gm.	[31.103 Gm.]
℥ 2 =	62.00 Gm	[62.207 Gm.]
℥ 4 =	125.00 Gm.	[124.414 Gm.]
℥ 8 =	250.00 Gm.	[248.823 Gm.]
℥ 16 =	500.00 Gm.	[497.656 Gm.]

2. *Of the determination of quantities of medicines by measure.*—This mode of determining quantities is resorted to by the pharmacist in the case of fluid medicines prescribed in terms of measure, and also, quite commonly, for the division into parts of gross amounts of solid medicines, such as pill-masses and powders. Thus, in the making of pills, a weighed quantity of the pill-mass is rolled out into a cylinder of definite length, and this cylinder is then cut into a number of equal subdivisions by a machine. Or, in the case of powders, the gross amount is carefully spread out into an elongated pile, which pile is then subdivided, by mensuration, into the proper number of parts. In the sick-room, the method of measuring has an important application in the determining of doses of fluid medicines—a purpose for which, as already seen, the process of weighing is practically inapplicable.

Mensuration, although, as stated above, less capable than weighing of yielding exact results, is yet, if done with proper precautions, quite accurate enough for the average requirements of dealings with medicines. But, to secure proper accuracy in the process, *regard must be had to the proportions of the measuring vessel.* For the peculiar physical fact must be remembered that the surface of a liquid in a vessel is concave if the vessel be partially filled, but convex if the same be brimful. Aligned, therefore, by a mark or brim, the volume of a fluid is not exactly in fact what it seems to be. And evidently the error will be proportioned to the relation between area of surface and volume. If the area of surface is small compared to the volume measured, the error will be small; if large, large. Accuracy in the measuring of volumes, therefore, depends vitally on the shape of the measuring vessel, the extreme of accuracy obtaining with the capillary-necked specific-gravity bottle, and of inaccuracy, let it be noted, with the shallow and flaring *spoon*. When anything like reasonable precision is required, therefore, tall and narrow instead of short and squat measuring vessels should be employed. For the measuring of medicines, tall, narrow graduates should be used for considerable volumes, and graduated pipettes for small. Even in the household, for the bedside measuring of doses (if the attendant be intelligent enough to use them), properly shaped graduates, or, for measures less than a fluidrachm (four cubic centimeters), graduated pipettes, should, in all cases requiring reasonable precision, be used in place of the faithless spoon.

The standard measures of capacity applied to dealings with medicines are, in the United States, *apothecaries'* or *wine* measure ; in England, *imperial* measure ; and, in countries where the metric scale is in vogue, *metric* measure, if volumetric methods be used at all. Domestically, the several capacities of the *drop*, the *spoon* in its various sizes, the *wine-glass*, and the *cup*, are also employed.

Apothecaries' measure is as follows :

TABLE OF APOTHECARIES' OR WINE MEASURE.

(Used in U. S. in the prescribing of medicines.)

Minim *Mi'ni-(mum).*	Fluidrachm *(Flui-dra'chma).*	Fluidounce *(Fluid-u'ncia).*	Pint *(Oc-ta'rius).*	Gallon *(Co'n-gius).*
♏	f ʒ	f ℥	O.	C.
60	=1			
480	8	=1		
7,680	128	16	=1	
61,440	1,024	128	8	=1

In reviewing this table, the following striking analogies appear between it and the table of apothecaries' weight. Both have denominations entitled *drachm* and *ounce*, respectively, and, in the system of measure, the numerical relation between the denominations *minim*, *fluidrachm*, and *fluidounce* are identical with those between the weight-denominations, *grain*, *drachm*, and *ounce*. The important differences between the two tables are, first, that in the table of measures there is no analogue of the weight-denomination, *scruple*; secondly, that the pint—the analogue among measures of the *pound* among weights—is of the value of sixteen of the next lower denomination, instead of twelve as is the case with the pound; and, thirdly, that there is, among measures, a denomination, the *gallon*, which has no analogue among weights. So far, however, as concerns the three lower denominations, which are those most used by the prescriber, the table of measures presents the convenient feature of an identity of inter-ratios with those obtaining between the analogous denominations in the scale of apothecaries' weight.

The various analogies in names and ratios noted above between the denominations of weight and measure, respectively, of the apothecaries' system, instinctively suggest some exact *correlation in fact* between those weights and measures, respectively, that are of similar title. But, unhappily, such desirable correlation does not exist. There is, however, an *approximate* correlation between apothecaries' weights and measures, through the medium *water*, which is close enough, in the majority of applications, for the purposes of the prescriber. A fluidounce of distilled water, at temperature 60° Fahr., weighs 455.7 grains—a number not so far from 480 but what, for the average of prescription-purposes, it is safe to reckon a fluidounce of water as weighing an ounce. By the same token it is commonly safe to regard a fluidrachm and a minim as weighing, respectively, a drachm and a grain.

Imperial measure—the system of measure in use in England—is shown by the following table:

TABLE OF IMPERIAL MEASURE.

(Used in England in the prescribing of medicines.)

Minim (*Mi'ni-mum*).	Fluidrachm (*Flui-dra'chma*).	Fluidounce (*Fluid-u'ncia*).	Pint (*Oc-ta'rius*).	Gallon (*Co'n-gius*).
min.	fldr.	floz.	O.	C.
60	=1			
480	8	=1		
9,600	160	20	=1	
76,800	1,280	160	8	=1

In this table, it is to be noted that the names of denominations are the same as in apothecaries' measure, and that, with the exception of the number of fluidounces to the pint, the denomination-relations are also identical. The actual values, however, even of the denominations of the same inter-ratio, differ slightly in the two systems, the minim, fluidrachm, and fluidounce of the imperial measure being but ninety-six per cent. of the respective capacities of the same denominations in apothecaries' measure. The imperial pint, however, being composed of a greater number of fluidounces than the apothecaries', is larger than the latter measure, and necessarily, therefore, the imperial gallon exceeds the gallon of the apothecaries' table.

Imperial measure has the advantage over apothecaries' of being exactly correlated to a system of weights, namely, the avoirdupois, which, as already seen, is, in part, the system of weights of the British Pharmacopœia. The correlation is, as usual, by the medium *water*, an imperial fluidounce of that liquid weighing exactly an ounce avoirdupois. Yet it must be carefully noted that in spite of this coincidence an imperial *minim* does not weigh exactly a *grain ;* this for the evident reason that an imperial *fluidounce* divides into 480 minims, while its equivalent, the avoirdupois *ounce*, contains but 437.5 grains.

Metric measure is shown in the table on page 73

but, as has been already indicated, this measure is rarely used in medical prescribing, the custom being, where the metric system is resorted to in prescription, to order everything, solids and fluids alike, by weight.

Concerning metric measure, the points of note are, first, that it shares with metric weight the conveniences inherent in a systematic decimal ratio between the denomination-values, and, secondly, that it is exactly correlated with this same metric weight through the usual fluid selected for such correlations, namely, distilled water. A milliliter (cubic centimeter) of distilled water at its greatest density weighs precisely a gramme, for the simple reason that, in devising metric weights, the weight of that measure of water was fixed upon to afford the standard unit of weight of the system. In the case of water, therefore, and, for prescription-purposes, in the case of fluids whose specific gravity is not far different from that of water, the gramme, in weight, and the cubic centimeter, in measure, are equivalents.

For a working table of equivalences between metric and apothecaries' measure, respectively, it will answer every practical purpose to apply the figures in the table of equivalences between metric and apothecaries' *weight*, respectively, already given, reading for the several titles *grains*, *drachms*, and *ounces*, the titles *minims*, *fluidrachms*, and *fluidounces*, respectively; and for grammes the title *cubic centimeters* (milliliters).

TABLE OF "METRIC," OR FRENCH DECIMAL MEASURES OF CAPACITY.

(Rarely used in the prescribing of medicines.)

Milliliter (Cubic centimeter). C.c.	Centiliter.	Deciliter.	Liter.	Decaliter.	Hectoliter.	Kiloliter.	Myrialiter.
10	=1						
100	10	=1					
1,000	100	10	=1				
10,000	1,000	100	10	=1			
100,000	10,000	1,000	100	10	=1		
1,000,000	100,000	10,000	1,000	100	10	=1	
10,000,000	1,000,000	100,000	10,000	1,000	100	10	=1

When metric measure is used in prescription-writing, the custom is, in notation, to write in ordinary decimal fashion, taking the denomination *cubic centimeters* for the integer, because of the equivalence of this measure to the gramme, the integer, in notation, of weights. Quantities by measure and by weight can thus, metrically, be written in column together, the suffixes "C.c." and "Gm." respectively indicating whether measure, in terms of cubic centimeters, or weight, in terms of grammes, is meant.

Of the measures *in domestic use* for dealing with medicines, the smallest is the time-honored *drop*. But the drop, though so commonly employed for the mensuration of medicines, is not, in the technical sense of the word, a measure at all, since its size differs enormously under different conditions, being affected not only by the viscosity of the fluid operated upon, but also, severally, by the shape, the surface-area, and even the position as regards degree of tip, of the dropping utensil. In general, large drops result in the case of fluids that are viscid, and of droppers that present, on the one hand, a large rather than a small surface from which the drop is to deliver, and, on the other, a concavity rather than a convexity for the fluid to cling to as it gathers in the pouring, preliminary to the fall in drops. And, by the same token, of course, comparatively small drops will be the rule in cases where these conditions, severally, are reversed As already hinted, an important factor in determining the size of the drop is, in the common case of the pouring, by drops, directly from a medicine-vial, the *degree of tip* of the vial at the time of the pouring. For when, as is the case when the vial is quite full, the contents begin to run out before the tip reaches the horizontal, the fluid running over the lip collects, before dropping, in the re-entrant angle formed by the projection of the lip from the neck of the bottle, and there, finding a concave nidus to cling to, does not fall until, comparatively, a goodly volume has accumulated. Hence, when the fall does occur, the drop is, comparatively, a large one.

In the case, on the other hand, of a half-empty vial, the fluid does not begin to run until the tip approaches the horizontal, under which circumstance, having only the narrow and convex edge of the lip itself to cling to, it falls, perforce, upon the gathering of a comparatively small quantity—falls, that is, in comparatively small drops. Hence it comes about that, even in the case of an identical bottle of medicine, the dose administered, if determined by dropping directly from the vial, will be very different in dimension according as the vial is full or half-empty on the occasion of the dropping.

Experiment shows, indeed, that a full bottle will yield, of an identical fluid, a drop more than half as large again as that resulting when the vial is half-empty. If, therefore, doses *must* be measured by drops, a pipette of standard size of orifice should be used as the dropping utensil instead of the lip of the vial. But by far the best plan, in the case of medicines whose dose is small, is as follows: Let the *aggregate* of a convenient number of doses be measured in some reliable way, as by use of a proper graduate or minim-pipette, and then let such aggregate be added to an identical number of spoonfuls of water actually measured with an individual spoon. If, then, the proportion, in volume, of medicine to water be, as it commonly will be under the conditions assumed, quite insignificant, a spoonful of the dilution, made in the manner described, will, *if measured by the same spoon originally used*, give, accurately enough for all practical purposes, the correct dose. This method has the further advantage that it is possible for the physician himself so to prepare the medicine, and thus personally to be assured that there will be no error, in dosage, in the administration.

A last point concerning the drop is the question

of its absolute dimension. Of course, from what has already been said, it is clear that no single dimension will represent this ever-varying quantity, but, so far as averages go, it is convenient to reckon on the equivalences, respectively, of a little over a minim for the drop of an aqueous fluid; from one-half to three-fourths of a minim for the drop of a tincture, a spirit, or a volatile oil, and a smaller proportion yet for that of an ethereal body. The drop in the case of *chloroform* is, because of the conjoint high tenuity and high specific gravity of this particular fluid, exceptionally small—as many as from 180 to 270, according to the conditions of the dropping, being required to fill the measure of a fluidrachm.

The *spoonful*, like the drop, is a very variable quantity, both on account of the faulty shape of the spoon-bowl for measuring purposes, and also because of the very variable size of spoons themselves, even of the same denomination. The spoon, therefore, should be limited, in use as a measure of capacity for medicines, to mixtures of comparatively indeterminate dosage.

In dimension, the average *teaspoonful* of the day will run six to the fluidounce, the *dessertspoonful* three to the fluidounce, and the *tablespoonful* three to two fluidounces, or six to four. In metric measure, the dimension of the *teaspoonful* is, on the average, five cubic centimeters, and that of the *tablespoonful* twenty cubic centimeters. In older times spoons were smaller than now, and the teaspoonful then rated equal to a fluidrachm, or four cubic centimeters, the dessertspoonful to two fluidrachms, and the tablespoonful to half a fluid-ounce, or sixteen cubic centimeters. Nowadays, however, it is safer to compute prescriptions, if doses are to be measured by the spoon, on the basis of the larger equivalents first given.

A *wineglassful*—a very vague term—is, as a measure, held to mean the capacity of the average sherry-glass, or about two fluidounces (sixty-two cubic centimeters). The *cupful* rates at from four to five fluidounces, and the *tumblerful* at from eight to ten or twelve, but all these vessels vary so in size as to be worthless as measures for any exact purpose.

CHAPTER V.

THE PRESCRIBING OF MEDICINES.

THE topic of the *prescribing* of medicines presents for technical study three distinct subjects, which, in logical sequence, are as follows: first, how to *compose* a prescription; secondly, how to *compute amounts* of ingredients; and thirdly, how to *write the document* in proper form and language.

I. *The Composing of a Prescription.*—Broadly considered, the subject of the composing of a prescription—meaning by the term the art of properly selecting the constituents of a composite medicament—embraces the consideration of the properties of the various articles of the materia medica in all their chemical, pharmaceutical, and physiological relations. Such consideration is, of course, not in place here, but there are yet certain general principles, bearing on the proper selection of the ingredients of prescriptions, whose discussion forms part of the technology of prescribing.

Proceeding to this discussion, the first point to be made is that a *given prescription should be com-*

posed with a single therapeutic aim only in mind. In other words, the rule should be followed that a given medicine should be for a single therapeutic purpose only, separate prescriptions being written and separate administrations being made under the circumstance that the patient stands in present need of medicinal attack from more than one point. Having, then, a single, definite therapeutic purpose in view, the next point is the consideration whether such purpose will be best fulfilled by a *single* drug of the proper category, or by a *team*, so to speak, of such drugs. No general rule can be laid down in this regard, since, as a matter of fact, sometimes the one condition obtains and sometimes the other. Thus, for example, even in the one class of remedies, cathartics, we find some, such as the purgative oils, which do their work best when given singly, while others, such as the resinous cathartics, operate more kindly, certainly, and often, too, more surely, by being prescribed in mutual association. The point in question, therefore, must be decided in each instance for the individual case, according to the circumstances of the same. The rule, however, should be observed *not to order*, in a given prescription, *a team of similar active constituents without a sound reason for so doing.* "Shot-gun" prescribing is to be condemned, if for no other reason than that it utterly defeats the exact clinical study of the therapeutic powers of drugs.

Assuming, then, that the active member, or team of members, of a prescription has been decided upon, the next consideration that presents is whether the medicinal action of the same can be rendered either, on the one hand, more *effective*, or, on the other, more *kindly*, by associating with the active medicine some drug of other quality. If such result can be so attained, then, of course, such association should be prescribed, on the principle of always aiming to secure, in the case of medicines, a maximum of therapeutic effect with a minimum of by-derangement. Now, as a matter of fact, it is quite often possible thus to enhance or to modify the medicinal action of a drug, so that a knowledge of such possibilities is essential to skilful prescribing. These modifications of the natural operation of a drug may result from a chemical or physical change wrought by another constituent of the prescription upon the active drug itself, or they may be effected by physiological impressions upon the subject. An example of the former kind is afforded by the action of a *solvent* in making more sure and speedy the absorption, and therefore the medicinal operation, of a salt; and of the latter, by the neutralization of the griping wrought by the rougher cathartics, by means of the peculiar physiological action of belladonna or of hyoscyamus, given in association with the purge.

Belonging to the category of additions to prescriptions for the purpose of making more *kindly* the operation of the medicine, are such additions as are designed, in the case of fluid mixtures, to *improve the taste.*

Obviously a medicine should be no nastier than need be, but, even apart from the question of general propriety, the matter of the taste of a medicinal mixture has most important practical bearings. For an unsavory potion may easily upset the sensitive stomach of an invalid—a result always deplorable, and, under certain circumstances, possibly serious also; and, in the case of children, an offensive dose leads to revolt and warfare, with, perhaps, disastrous results to patient and physician, both. For these reasons, the making of a mixture to be as pleasing as possible to eye, nose, and palate, all, is really an element of the art of prescribing of prime importance, and the young prescriber is, therefore, earnestly advised to make the same a subject of serious study.

Obviously these remarks apply to *fluid* medicines: if the stuff be solid, the method of taking by capsule, wafer, or pill will so conceal taste as to make the same a matter of no moment. Now, the things to add to fluid medicines to cover taste are mainly *sugar* or *syrup*, or preparations from pleasantly flavored aromatics—viz., the syrups, waters, and spirits derived from those drugs, or a minute dash of their essential oils. Lists of the pleasantly flavored waters, spirits, and syrups of the Pharmacopœia were purposely detailed when speaking of forms of medicines. A judicious use of these flavoring agents may not only make a potion less nasty, but may prevent its sickening, and so have a really important influence on the therapeutics of the active drug.

A third consideration, in the selection of the members of a prescription, relates to such constituents as may be necessary to *effect solution*, to *afford volume*, or, as in the case of the excipient of a pill-mass, to *determine form*. Evidently the selection of substances for these several purposes can follow no fixed scheme, but must be determined in each instance by the considerations affecting the individual case.

On review, it thus appears that a prescription may, with propriety and possible advantage, comprise constituents for the fulfilling of the three separate aims, respectively, of, first, producing the desired *medicinal impression;* secondly, improving the *quality* of such impression by either enhancement, modification, or both; and, thirdly, giving to the prescribed body the proper *condition*, *volume*, or *form*. To give, then, technical names to the several members of a prescription, according to their respective purposes, we dub the member, or team of members, that is to do the medicinal work, the *basis;* that which enhances the working of the basis, the *adjuvant;* that which corrects some disagreeable by-effect of the same, the *corrigent*, or *corrective;* and that which gives volume or form, the *vehicle* or the *excipient*, respectively.

A general consideration affecting the selection of the members of a prescription is the point that regard must be had to the chemical and chemico-physical properties of the substances proposed for association, lest, inadvertently, untoward reactions be determined. For things which, from the purely therapeutic point of view, might seem well fitted for combination in the same prescription, may yet easily be of such chemical tendencies as to develop, upon admixture, unsightly, inert, or even explosive compounds. The *chemistry* of medicinal substances has, therefore, an important practical bearing on correct prescribing, and in this place it is proper to discuss, in a general way, those chemical reactions whose effects most need to be taken into present consideration. Such are as follows:

1. *Acids and bases tend to combine, forming salts.* This reaction may be utilized in order to *get* some salt that may be needed; but, if *free acidity or alkalinity* be aimed at, acids and bases must not be prescribed together.

2. *Strong acids or bases generally displace their own weaker brethren when met with in saline combination.* Here the word "generally" is used advisedly, for, under the circumstance that an *insoluble* salt will result, the reverse may obtain, and, through chemistry's imperious passion for precipitates, a weaker acid or base may displace a stronger. Ordinarily, however, the fact is as stated, and its bearing on prescription-combinations is obvious.

3. *Salts in solution exchange radicals, or acids or bases displace their brethren in saline combination, if, thereby, an insoluble compound can be formed.* This is a fact in chemistry that quite generally obtains, and whose bearing on prescribing is important. In the first place, the reaction may be a convenience, of which we actually avail ourselves as a means of getting a thing that we happen to want in condition of precipitate. The well-known *black* and *yellow* washes of mercury are examples in point. In the second place, the reaction may *make no practical difference* in either medicinal activity or other qualities of the com-

bination prescribed. This fact is often overlooked by systematists, and combinations are solemnly warned against as "incompatibles," solely because a precipitate occurs in the compounding, the admonisher forgetting that the chemical activities of the alimentary apparatus will often dispose of a precipitated substance as quickly as of one in actual solution.

In the third place, however, the reaction *may* make a most important difference, either because the precipitate is difficult of solution by the digestive fluids, and therefore is medicinally feeble or inert, or because the presence of the precipitate, as such, in the mixture is unsightly, or makes the same awkward or dangerous for administration. The more prominent of mutually precipitant solutions are shown in the table on page 87. Here the precipitates accredited to solutions of salts of the alkaloids and the metals occur with the generality of those bodies, though not absolutely with all.

4. *Things in solution precipitate on addition of excess of a fluid in which they are insoluble.* This fact is of concern principally in the matter of *aqueous* and *alcoholic* solutions. Water and alcohol are the main pharmaceutical solvents; many things dissolve in both fluids, but many others in but one. When prescribing any of the latter category, therefore, a considerable admixture of

TABLE SHOWING NOTABLE MUTUALLY PRECIPITANT SOLUTIONS.

	Solutions of *Alkalies.*	*Carbonic acid* and solutions of *carbonates.*	*Sulphuric acid* and solutions of *sulphates.*	*Phosphoric acid* and solutions of *phosphates.*	*Boric acid* and solutions of *borates.*	*Hydrochloric acid* and solutions of *chlorides.*	*Hydrobromic acid* and solutions of *bromides.*	*Hydriodic acid* and solutions of *iodides.*	Solutions of *sulphides.*	*Tannic acid.*	*Arsenical* solutions.	*Albumen.*
Alkaloidal solutions (generally)	prec.	prec.		prec.	prec.			prec.		prec.		
Metallic solutions (generally)	prec.	prec.		prec.	prec.				prec.	prec.	prec.	prec.
Lead solutions...	prec.	prec.	prec.	prec.	prec.	prec.	prec.	prec.	prec.	prec.	prec.	prec.
Silver solutions...	prec.	prec.	prec.	prec.	prec.	prec.	prec.	prec.	prec.	prec.	prec.	prec.
Calcic solutions...	prec.	prec.	prec.	prec.								
Magnesic solutions	prec.	prec.		prec.								
Albuminous solutions..........										prec.		
Gelatinous solutions............										prec.		

the non-solvent fluid must be avoided. And the list to be borne in mind is that *albuminous*, *gelatinous*, *gummy*, and *saccharine* bodies, and many *salts*, tend, as a class, to dissolve in *water*, but not much in *alcohol;* while, on the other hand, *volatile oils* and *resins*, including balsams and camphor, tend to dissolve in alcohol, and but slightly or not at all in water.

5. *Powerful oxidizing agents may determine explosions on concentrated admixture with readily oxidizable substances.* The medicinally-used powerful oxidizers are *chromic* and (strong) *nitric* or *nitrohydrochloric acids*, and *potassium chlorate* and *permanganate*, and the most easily combustible bodies are *oils*, *alcohols*, and *ethers* (including among the alcohols *glycerin* and *sugars*, which chemically belong to the alcohol genus), *dry organic* substances generally, *sulphur* and *phosphorus.* Not all of these combustibles will explode on treatment with all of the oxidizers, but detail is unnecessary, for it is just as well, in practice, to avoid any combination of the one class of bodies with the other.

Such, then, are the reactions determining incompatibility which affect considerable numbers of medicinal substances; other prominent instances of incompatibility are individual, and are best studied and thought of in connection with the individual substances concerned. Many things, even such as corrosive sublimate or syrup of ferrous iodide, are, chemically, so very vulnerable that the practical rule obtains to associate with them, in prescribing, nothing but plain water.

Passing now from the principles affecting the composing of a medicinal combination, we should logically next discuss the art of computing amounts. Inasmuch, however, as in actual practice it is the custom—and a wise one—to write down the titles of all the constituents of a prescription before proceeding to calculate quantities, we shall find it more convenient, in study, to follow the order thus suggested, and to leave, therefore, the consideration of the computation of amounts to the last. We proceed, then, at once to the twofold topic of the *form* and *language*, or, to phrase it simply, the *expression* of a prescription.

II. *The Expression of a Prescription.*—In *form* a prescription begins with the *name of the patient* for whom the medicine is ordered; then follow the *directions to the pharmacist* of what ingredients to take and how to compound them; next, a transcription of the desired *labelling* as to dose and frequency of giving; fourthly, *date* and *signature* of the author; and fifthly, any special order concerning the prescription itself, such as "not to be renewed," "to be returned," "not to be shown to the patient," etc. Such orders, not being an integral part of the prescription proper, may be put, indifferently, at the top or bottom of the paper.

Now, concerning these several parts of a prescription, the following general points may first be made. The *name of the patient* should always be entered upon the prescription as a safeguard against a possible mistake—on the one hand, of the dispenser, in delivering the wrong bottle to the customer, or, on the other, of the nurse in administering the wrong medicine to the patient, supposing, as may happen, that two or more patients are at the same time under the same care. The *directions for compounding*, which follow next, should be written *deliberately* and *thoughtfully*, under the consciousness that a "slip of the pen" may cost a human life; and *fully* and *legibly*, with a realization that a *slip of the eye* on the part of the pharmacist may lead to a like result. In setting down the titles of the ingredients, therefore, no abbreviations should be practised further than possibly to lop off the case-endings (Latin) of nouns and adjectives; but far better is it, if the prescriber's latinity will stand the strain, not to do even this, but to write in good readable hand the title in its entirety. The third part of the paper, the *directions for the taking*, which the dispenser is to transcribe upon the label of the package, should, for the same reason of seeking all possible surety against mistakes, be written fully, intelligibly, and, of course, again, legibly. The empty phrase, "use as directed," so often senselessly ordered to be entered in lieu of the directions themselves, is as solemnly absurd as would be a legal contract reading "we hereby agree to do as we have agreed"; and, of course, amounts to letting the bottle be launched, label-less, upon its errand, to work weal or woe, according as human forgetfulness and misunderstanding may chance to determine. The fourth feature is *date* and *signature* of the prescriber—a feature that should as invariably appear in a prescription as in any other document having a business-bearing. Fifthly, the special directions concerning disposal of the prescription, considered as a piece of property of its author, are, of course, at the same author's discretion.

So much in a general way. And passing now to matters of detail, we need discuss only our second part of the prescription, namely, that which constitutes its body—the directions for the compounding of the medicine. This is, in form, a straightforward order telling the compounder to take such-and-such things and do so-and-so with them. It is commonly expressed, in style and phrase, according to this general formula:

> Take, *x* quantity of *A*,
> *y* quantity of *B*,
> *z* quantity of *C*, etc.
> Do so-and-so [with them].
> Label [the package]: "........"

In setting down the titles of the ingredients, the natural order is followed of writing, first, for the *basis*, or series of bases, next for the *adjuvant*, thirdly for the *corrigent*, and, fourthly, for the *vehicle*—flavoring agent first, inert diluent last.

Disposing thus of *form*, we pass to the important matter of *language* of the prescription. So far as concerns the entering of the name of the patient, the writing out of the directions for administration for copy upon the label, the dating and signing, and the enjoining of special observance about the prescription—all these parts are, in the United States, quite generally, and entirely properly, written in the vernacular. The part, however, which comprises all *directions to the pharmacist*—the part included in the foregoing formula between the words "take" and "label," inclusively—is, by an ancient custom whose wisdom it is beside the mark to discuss, almost universally written—or supposed to be written (!)—

in Latin. A certain knowledge of the Latin language thus presents itself as a necessity to the prescriber, and to the medical student unfamiliar with that tongue looms a very mountain of untold terrors, whose pathways he despairingly makes no attempt to tread. But be it ours to reassure the faint-hearted traveller—the difficulties are neither so many nor so grave as he fears. For apart from the vocabulary of titles of medicines, which, of course, must be learned by rote, the words and phrases of prescription-usage are few, and the forms so set that a very little of etymology and syntax suffices for proper rendering. The student with no knowledge of Latin is, therefore, earnestly urged to master so much of the vocabulary and grammar of that tongue as may save him from disgrace before latinist patient or pharmacist on each occasion of his issuing a prescription. So much of Latin, then, if to be learned, must be taught, and to such teaching we will without further apology betake ourselves.

Reverting to our formula, we brush away with one sweep, so far as latinizing is concerned, fully one-half the wording, by the custom of expressing invariable or oft-recurring words or phrases by abbreviations or arbitrary symbols. Thus, as follows: the verb "take," which begins the formula, is expressed by the symbol "℞," whose curious form is, in part, the initial letter of the Latin word *recipe*, "take [thou]," and in part the astronomical sign of the planet Jupiter, "♃," formerly used to symbolize a prayer to the deity Jupiter for divine blessing upon the remedy to be ordered. Next, the words concerned in *expressing quantity* are also invariably symbolized, in using the apothecaries' system, by the employment of the established denomination-symbols, and the expressing of numerals, Roman fashion, by the small letters of the alphabet written *after* such symbol; and, in the case of the metric system, by the use of ordinary Arabic notation, in decimal style, with the abbreviation "Gm." for *gramme*, or "C.c." for *cubic centimeter*, *following* the numeral. Lastly, the word "label" is expressed by the abbreviation "S.," being the initial letter of the Latin word *signa*, "mark [thou]."

We are thus happily narrowed down, for translation into Latin, to the enumeration of the titles of ingredients, on the one hand, and to the phrases containing the directions for compounding, on the other. The latter division, being the simpler, will be first considered. In the majority of cases, the directions for compounding are the simplest possible, consisting merely of a single word, such as *mix*, or *dissolve*. So frequently indeed, does the word "mix" occur as the entirety of the order, that it too is commonly symbolized, being expressed by the abbreviation "M.," the initial of the word *misce*, "mix [thou]." For the rest, the commonest occurring phrases can be correctly latinized by the use of the following vocabulary in connection with a previous knowledge of the technical titles of forms of medicinal preparations, and by the application of the usages of Latin etymology and syntax shortly to be expounded. Should, however, any pharmaceutical procedure require to be detailed in prescription, too complicated for the latinizing of its statement by the aids thus offered, then let the prescriber quietly "drop into" English for that sentence. Not only would this doing be permissible, but it would, in this country, be even advisable, lest unusual Latin "stump" the *pharmacist*, to the confusion of prompt and faithful compounding.

TABLE SHOWING ODD WORDS OCCURRING IN PRESCRIPTION-PHRASES.

I. *Verbs, imperative*, object to be in the accusative case (analogue of English "objective").

A'dde, add.	*Fac*, make.	*Re'cipe*, take.
Co'la, strain.	*Fi'ltra*, filter.	*Si'gna*, mark.
Di'vide, divide.	*Ma'cera*, macerate.	*So'lve*, dissolve
Exte'nde, spread.	*Mi'sce*, mix.	*Te're*, rub.

II. *Verbs, subjunctive*, taking subject or predicate nominative (analogue of the English nominative).

Bu'lliat, let [it] boil.

Fi'at (singular), *Fi'ant* (plural), let [it, them] be made [into].

III. *Verbal adjective* (participle), to agree with its noun in gender, number, and case.

Divide'nd-us (masculine), *-a* (feminine), *-um* (neuter), to be divided.

IV. *Prepositions*, following noun to be in the accusative case:

Ad, to, up to. | *In*, into. | *Supra*, upon.

V. *Prepositions*, following noun to be in the ablative case:

Cum, with. | *Pro*, for.

VI. *Miscellaneous words and phrases:*

A'na, of each (Greek).
Be'ne, well.
Bis, twice.
De'in, thereupon.
Et, and.
Grada'tim, gradually.
Gutta'tim, by drops.
Non, not.
Se'mel, once.
Si'mul, together.
Sta'tim, at once.
Ter, thrice.

Ad satura'ndum, to saturation.
Nu'mero, to the number of.
Qua'ntum suffi'ciat, as much as may be necessary.
Pro re na'tâ, according to need.

So we come now to the only part of the prescription whose correct translation into Latin gives any serious trouble, namely, the enumeration of the medicinal things which the compounder is to take. Here, so far as mere vocabulary is concerned, the words are, of course, numerous, and the what means what and the correct spelling thereof are things that must, of necessity, be learned by rote, by hard "digging." But the technicalities with which we have here to deal are those of the how properly to fit our words together to mean what we want to say with them. We may find in the dictionary that *compositus* means "compound," *extractum* means "extract," and *colocynthis* "colocynth," but yet how shall we say "compound extract of colocynth," in good Latin, in ordinary statement, on the one hand, and in the special prescription-phrase to "take *x* quantity *of* 'compound extract of colocynth'" on the other? Again let the novice take heart; though the words are many, the amount of etymology and syntax required for their proper setting is not more than a bright mind can grasp, in the principle, in an hour. Then a little daily practice with the tables to be exhibited will soon make of a willing apprentice a good *terminological* expert!

Let us now take a few examples of medicine-titles and analyze their construction:

Compound extract of colocynth.
Quinine sulphate (meaning sulphate *of* quinine).
Powder of opium.
Liniment of soft soap.
Wine of root of colchicum.
Mercury with chalk.

Here we find that we have to deal with *nouns* and, occasionally, *adjectives;* that the leading noun of a title is in the *nominative case*, but that the dependents are in the *objective*, most of them following the preposition "of," although one ("chalk") follows the preposition "with." Let us see now how the same will appear in prescription-phrase:

"Take, *x* quantity *of* compound extract of colocynth,
y quantity *of* sulphate of quinine," etc.

Evidently here the dependent nouns, such as "colocynth" and "quinine," remain unaffected in relation, but evidently, on the other hand, the relation of the leading noun is, in each case, changed. *It* now becomes a dependent upon the preceding noun "*quantity*," and so appears, like its own dependent, in the objective case after "of," instead of being, as in ordinary arbitrary statement, in the nominative. Next let us consider certain prescription-forms occasionally arising:

"Take, *x* quantity of oil of castor,
The yolk of one egg."

Or again:

"Take, *x* quantity of A,
y quantity of B,
water, as much as needed to make the mixture measure *z* quantity."

In both of these examples we find that the last entry does not read to take a given *quantity of* the thing, but to *take the thing itself*—the *yolk* in the one case and *water* in the other. Here, then, the words "yolk" and "water" are not in the objective after *of*, but in the objective without a preposition, as the immediate "objects" of the verb *take*.

Our analysis of English titles shows us, thus, that we need, in the way of grammar, to know how, given a nominative, to form the expression for: (1) objective case after *of;* (2) objective case after *with;* (3) objective case following a verb. Now, in Latin, case is expressed, as in the English "possessive," by modification of the *ending* of the word, and, in Latin, each of the conditions of case cited above constitutes an individual "case," expressed by individual ending. Our examples, then, illustrate Latin "cases," as follows: English *nominative* equivalent to Latin *nominative;* English *objective after "of,"* equivalent to Latin *genitive;* English *objective after "with,"* equivalent to Latin *ablative;* English *objective following verb*, equivalent to Latin *accusative.* In Latin there are also two other cases, *dative* and *vocative,* but these do not occur in prescription-writing.

The whole technicality, then, of properly framing medicine-titles in Latin, and of setting down such titles under the syntax-conditions found in prescription-phrasing, resolves itself into the correct changing of ending of the dictionary-word to fit expression of *case.* But, unfortunately, this is by no means the simple affair it is in English. In English, the whole matter is a simple modification of ending for a single case (possessive) of nouns, and no modification whatever for adjectives; but, in Latin, each of the several cases has (generally) a special ending, and this for adjectives and nouns both; there are, furthermore, five independent schemes of forming such case-endings, and in the case of adjectives the scheme varies, according to the *gender* of the noun which the adjective affects. Lastly, these same genders are wholly arbitrary, depending more on peculiarity of nominative ending of a noun than on any "gender," in the proper sense of the word, real or allegorical, which the thing signified by the noun may possess! Here, then, is complexity, and in this complexity resides the whole practical difficulty of the latinizing of prescriptions. To thread the labyrinth thus presented, we see that we need to know three separate things: *First*, the several schemes of forming case-endings—the *declensions*, in short, in order to get always the proper inflection for a given "case"; *secondly*, the system by which, with a given noun in the nominative, we are to recognize the declension to which such noun belongs; and, *thirdly*, the system by which we are to know *genders*, in order to tell, with a given noun, by what declension and subdivision thereof to inflect any adjective we may propose to attach to the same.

First, of the *declensions*, or schemes of forming case-endings. As already said, there are five such, but of these one—the fifth—offers, in ordinary prescription-vocabulary, but a single example, viz., the ablative case *re*, in the phrase "pro re natâ," meaning, in free translation, "as necessity arises," referring to indications for dosing. We need, then, to know but four declensions, and, of these, only the inflections for the cases concerned in prescription-writing, namely: in the singular number, the *nominative*, *genitive*, *accusative*, and *ablative;* and, in the plural number, *nominative*, *genitive*, and *accusative*. Following is a table of such parts:

TABLE OF PARTS OF DECLENSIONS
Concerned in Prescription-Latin.
Nouns and Adjectives.

	FIRST DECLENSION. *f.*		SECOND DECLENSION. *m.*	*n.*	THIRD DECLENSION. *m. and f.*	*n.*	FOURTH DECLENSION. *m.*
				Singular Number.			
Nominative..	-a	(e)	-us(-os)	-um(-on)	(various)	(various)	-us
Genitive	-æ	(es)	-i		-is		-ûs
Accusative..	-am	(en)	-um (-on)		-em	(like nom.)	-um
Ablative	-â		-o		-e		
				Plural Number.			
Nominative..	-æ		-i	-a	-es	-a	-us
Genitive	-arum		-orum		-um, -ium		-uum
Accusative..	-as		-os	-a	-es	-a	-us

(Fifth declension exemplified only in ablative singular "re" in phrase *pro re natâ.*)

	Cardinal Numerals.							
	m.	*f.*	*n.*	*m.*	*f.*	*n.*	*m. and f.*	*n.*
Nominative....	un-us	-a	-um	du-o	-æ	-o	tr-es	-ia
Genitive		-ius		-orum	-arum	-orum	-ium	
Accusative....	-um	-am	-um	-os	-as	-o	-es	-ia

INDECLINABLE.

Alcohol,	Elixir,	Naphtol,
Amyl,	Eucalyptol,	Pyrogallol,
Buchu.	Kamala,	Salol,
Cajuputi,	Kino,	Sassafras,
Catechu,	Matico,	Sumbul,
Chloral,	Menthol,	Thymol.
Cusso,	Methyl,	

And cardinal numerals signifying a higher number than *three.*

In this table, the endings shown are but excerpts from the total variety of endings which the declensions afford, endings not exemplified in prescription-terms being purposely omitted. The latinist will, therefore, miss from the table nominatives of second declension in *er*, and of fourth in *u;* ablative singular of third declension in *i*, and neuter plurals in *ia*, etc.; the object being, in the table as in the text, to restrict the teaching to only so much of latinity as is actually applied in prescription-writing. In the table, the *genders* of nouns of different endings, so far as they are determinable by the ending, are expressed by the abbreviations *m.*, *f.*, and *n.*, meaning *masculine*, *feminine*, and *neuter*, respectively; and the endings enclosed in *parenthesis* are those of Greek nouns which have been adopted into Latin in somewhat of their original Greek dress.

In applying this table the case-endings of nouns and adjectives are appended to the so-called *root* or *stem* of the word, which, in words of the first, second, and fourth declensions, is found easily enough by subtracting from the nominative the case-ending, but which in the *third* declension is often, unfortunately, itself so modified or curtailed in the nominative that we must pass to the genitive for its revelation.

Thus the oblique cases *cantharidis*, *canthari-*

dem, *cantharide*, *cantharides*, show plainly the root or stem "cantharid-" while, yet, the nominative is *cantharis*, a form whose dissection fails to show the root. The third declension thus has this peculiar disability: that, knowing the nominative, we cannot therefrom always deduce the genitive, and hence must, in the case of members of this declension, do the double memorizing of nominative and genitive. The declension of *cardinal numerals* is irregular in the case of *unus*, "one," and *duo*, "two," but *tres*, "three," takes the regular endings of the third declension, plural. All other cardinal numerals are indeclinable, as are also, as the table shows, certain words of "barbarous" origin, adopted into Latin without a Latin dress, and hence necessarily exempted from Latin inflections. With all indeclinable words there is no change, in any case, from the spelling of the nominative.

Next, as regards *genders*. Gender of nouns is in Latin determined by two independent considerations, one being the nature of the thing signified, and the other the ending and declension of the noun; and of these considerations, if they conflict, the former takes precedence. That is, if a noun by its declension-ending ought to be masculine, but yet signifies a thing regarded in Latin idiom as intrinsically feminine, that noun will be feminine, in defiance of declension-ruling. Of the kinds of things which, by their class-nature, thus determine gender, there is but one exemplified among prescription-nouns, namely, *trees*, the names of which, by a Latin custom, furnish *feminine* nouns, no matter what the declension-endings. Yet here the point must be noted that the rule obtains only in *ancient Latin current tree-names*, such as *quercus*, "oak," *ulmus*, "elm," etc., the technical Latin botanical names of modern invention, signifying trees, taking the natural gender of their declension-ending. Thus while, as just cited, *ulmus* is feminine, *eucalyptus* is masculine. This whole matter, however, hardly merits the time it takes for its statement, since the number of tree-name nouns of prescription-occurrence that are thus forced to be feminine, when they ought to be of other gender, are very few, and of them but four are, in drug-titles, followed by an adjective, so as to require attention to be paid to the gender at all. The full list of such classical tree-names occurring in pharmacopœial titles is: of the second declension, *juniperus*, *prunus*, *rhamnus*, *sambucus*, and *ulmus*; of the third, *rhus* (in the title "rhus glabra"); and, of the fourth, *quercus*. And of this list the four taking adjectives in drug-nomenclature are *prunus* with adjective *Virgini-*

ana, *quercus* with adjective *alba*, *rhamnus* with adjective *Purshiana*, and *rhus* with adjective *glabra*.

Disposing thus of the bearing of signification upon gender, for the rest the genders of nouns are determined by declension-ending, and a study of our declension-table shows that for all declensions, except the troublesome *third*, again, each form of nominative case-ending carries with it a special gender. Thus all (prescription-used) nouns of the first declension are feminine, and those of the fourth (excepting, of course, the tree-name nouns just cited) masculine; while in the second declension all nouns in *-us* (certain tree-names excepted again) are masculine, and in *-um*, neuter. As to third-declension nouns, the various endings of this declension give all genders, and, unfortunately, all nouns of the same ending are not always of the same gender. A little hard memorizing thus becomes necessary here, but only a little, for happily in very many instances all nouns of a certain ending and genitive-formation take the same gender, and, with the majority of those whose genders are arbitrary, prescription-usage does not require the gender to be known. The most prominent instance of a natural group of nouns of the same gender is in the case of nouns in *-as* forming genitive in *-atis*, and nouns in *-is*, genitive *-itis*, such as *sulphas* and *sulphis* —nouns all signifying chemical genus among salts. These are all of the same gender, which, after many vicissitudes in different Pharmacopœias, settled down, in the sixth revision of our own authority, into masculine. Lastly, as to indeclinable nouns, these are all neuter.

For the *determining of declension and gender* of declinable nouns, a table is convenient that shall show, for each nominative singular ending, the declension and gender of such Latin nouns as occur in pharmacopœial medicinal titles and common prescription-use. Such a table will be found on pages 110–114. In this table, all nouns of the third declension—the one so troublesome because of diversity of genitives and genders—are catalogued in full under their several nominative-endings, and the genitives given in parenthesis. The genders also are given throughout for completeness' sake, although, as already said, genders need to be known only in the minority of instances. Genders are designated by the abbreviations *m.*, *f.*, and *n.*, signifying *masculine*, *feminine*, and *neuter*, respectively.

Adjectives are declined after the manner of nouns, and, in Latin, agree with their respective nouns in gender, number, and case. A table showing the schemes of declension of adjectives concerned in prescription-writing will be found on page 115.

Concerning the scheme shown in the table we may note the following: Scheme I. embraces the very great majority of adjectives, and the neuter ending *-on* instead of the usual *-um* occurs in but a single example, *diachylon.* Scheme II. comes next in order of membership, such adjectives as *mitis*, "mild," *mollis*, "soft," belonging to this family. In the neuter of this scheme we have examples of words in *-e* of the third declension—words not occurring among pharmacopœial nouns. Scheme III. is a peculiar scheme for declining the "comparative" of certain adjectives, and presents for us but a single example, *fortior* (masc. and fem.), meaning "stronger." Scheme IV. has, in our present Pharmacopœia, but two examples in *-ens*, *effervescens* and *recens*. Viewing the schemes together, we see that the nominative-ending carries with it the showing of declension and gender in all cases.

Table Showing Declension and Gender of Latin Nouns Occurring in Titles of U. S. Pharmacopœial Medicines and in Common Prescription-Terms.

Nominative Singular ending in **-a** :

All First Declension and Feminine, except Kamala, indeclinable, *and* (of Greek origin) *the following in* **-ma** :

Physosti'gma (physosti'gmatis), 3*d*, *n*.	E'nema (ene'matis), 3*d*, *n*.
Aspidospe'rma (aspidospe'rmatis), 3*d*, *n*.	Catapla'sma (catapla'smatis), 3*d*, *n*.
	Gargari'sma (gargari'smatis), 3*d*, *n*.
	Theobro'ma (theobro'matis), 3*d*, *n*.

Nominative Singular ending in **-us** :

All Second Declension, Masculine, except—

Juni'perus, 2*d*, *f*.	Rhus (rho'is), 3*d*, *f*. ("rhus glabra")
Pru'nus, "	Fru'ctus, 4*th*, *m*.
Rha'mnus, "	Spi'ritus, "
Sambu'cus, "	Que'rcus, 4*th*, *f*.
U'lmus, "	

Nominative Singular ending in **-os**:

Comprise only the following—

Flos (flo′ris), 3*d*, *m*.	Bos (bo′vis), 3*d*, *m*. or *f*.

Nominative Singular ending in **-um**:

All Second Declension, Neuter.

Nominative Singular ending in **-on**:

Comprise only the following—

Eriodi′ctyon, 2*d*, *n*.	Eri′geron (erigero′ntis), 3*d*, *n*.
Hæmato′xylon, "	Li′mon (limo′nis), 3*d*, *m*.
Toxicode′ndron, "	

Nouns of all other endings are of the Third Declension, and are as follows:

Ending in **-c**:

Lac (la′ctis), *n*.

TABLE SHOWING DECLENSION AND GENDER OF NOUNS (*Continued*).

Ending in **-el** :

Fel (fe'llis), *n.*	Mel (me'llis), *n.*

Ending in **-en** :

Alu'men (alu'minis), *n.*	Se'men (se'minis), *n.*

Ending in **-o** :

(*-io*)	(*-ago*)
Confe'ctio (confectio'nis), *f.*	Mucila'go (mucila'ginis), *f.*
Lo'tio (lotio'nis), *f.*	(*-bo* and *-po*)
Po'rtio (portio'nis), *f.*	Ca'rbo (carbo'nis), *m.*
Tritura'tio (trituratio'nis), *f.*	Pe'po (pepo'nis), *m.*
	Sa'po (sapo'nis), *m.*

Ending in **-r**:

8

(*-er*)

Æ′ther (æ′theris), *m.*
Pi′per (pi′peris), *n.*
Zi′ngiber (zingi′beris), *n.*

(*-or*)

Li′quor (li′quoris), *m.*

(*-ur*)

Su′lphur (su′lphuris), *n.*

Ending in **-s** :

(*-as*, genitive *-atis*)

Ace′tas (aceta′tis), *m.*
[*and all salt names in -as*].

(*-as*, genitive *-adis*)

Ascle′pias (asclepi′adis), *f.*

(*-is*, genitive *-itis*)

A′rsenis (arseni′tis), *m.*
[*and all salt names in -is*].

(*is*, genitive *-eris*)

Pu′lvis (pu′lveris), *m.*

(*-is*, genitive *-is*)

Ca′nnabis (ca′nnabis), *f.*
Digita′lis (digita′lis), *f.*
Hydra′stis (hydra′stis), *f.*
Sina′pis (sina′pis), *f.*

(*-os*, *see ante*)
(*-us*, *see ante*)

Table Showing Declension and Gender of Nouns (*Continued*).

Ending in **-s** (*continued*) :

(*-is*, genitive *-idis*)

A′nthemis (anthe′midis), *f.*
Ca′ntharis (cantha′ridis), *f.*
Colocy′nthis (colocy′nthidis), *f.*
Hamame′lis (hamame′lidis), *f.*
I′ris (i′ridis), *f.*
Ma′cis (ma′cidis), *f.*

(*-ns*)

Ju′glans (jugla′ndis), *f.*

(*-ps*)

A′deps (a′dipis), *m.*

(*-rs*)

Pars (pa′rtis), *f.*

Ending in **-x** :

(*-ax*)

Bo′rax (bora′cis), *m.*
Sty′rax (styra′cis), *m.*

(*-ex*)

Co′rtex (co′rticis), *m.* and *f.*
Ru′mex (ru′micis), *f.*

(*-ix*)

Pix (pi′cis), *f.*
Ra′dix (radi′cis), *f.*

(*-ux*)

Nux (nu′cis), *f.*

(*-lx*)

Calx (ca′lcis), *f.*

TABLE

Showing Schemes of Declension and Gender of Adjectives occurring in Pharmacopœial Medicinal Titles and in Prescription-Phrases.

SCHEME I.—*Second and First Declensions combined.*

Masculine.	*Feminine.*	*Neuter.*
-*us* [2d dec.]	-*a* [1st dec.]	-*um* (-*on*) [2d dec.]

SCHEME II.—*Third Declension.*

Masculine and Feminine.	*Neuter.*
-*is* (genitive -*is*).	-*e* (genitive -*is*).

SCHEME III.—*Third Declension.*

Masculine and Feminine.

-*or* (genitive -*oris*).

SCHEME IV.—*Third Declension.*

All Genders.

-*ens* (genitive singular -*entis*) ; (genitive plural -*entium*).

A special and last point in the latinity of prescription-writing is that Latin idiom imposes a different *order of words* in the sentence from what is the custom in English. Referring to our original prescription-formula, we see that the wording is, *take x quantity of A*. Now this, in Latin, would read, *take, of A, quantity x*, and such Latin order, as is well known, is observed in prescription-writing. Again, such a title as *leaves of belladonna* would, in Latin, have the words reversed in order, reading, of *belladonna the leaves*. Such Latin order is commonly followed in the latinizing of titles of medicines, except that, in the case of pharmaceutical preparations, the word signifying the kind of preparation—tincture, extract, etc.—precedes its dependent, as in English. Hence we have the incongruity of the titles, *opii pulvis*, "powder of opium"—a *condition* of opium; but *tinctura opii*, "tincture of opium"—a *preparation* made from the drug. Another point is that, in Latin, adjectives follow the nouns they affect, instead of preceding them, as in English; and this idiom is commonly observed in pharmaceutical Latin. *Soft soap* is, therefore, *sapo mollis; mild chloride* is *chloridum mite*, etc.

So much, then, for the twin subjects *form* and *language* of a prescription; and since the technicalities here are arbitrary, it is best not to rest with their mere general exposition, but to fix our newly acquired knowledge by working out a few examples. This we will do, then, in this place, leaving the matter of *quantities* for future consideration.

We wish to order for Mrs. A. B. a stomach-bitter, and we select *quinine sulphate.* Forthwith, then, we set down the phrase "For Mrs. A. B.," and follow it with the order "take of sulphate of quinine"—as yet not fixing the amount. Having got thus far, we bethink us in what pharmaceutical form this quinine salt shall be given, and we determine upon the fluid form, and that the salt shall be in actual solution. Then occurs the chemical point that quinine sulphate needs the help of an acid to dissolve it in ordinary fluids, and so perforce we must add an acid to our prescription; we elect to take *aromatic sulphuric acid,* and so write next the words [take] "of aromatic sulphuric acid"—a quantity to be determined by the amount of quinine. Next we turn our thoughts to the ingredients to make up the fluid vehicle in which the quinine salt is to be dissolved. This is to be essentially aqueous, but it occurs to us as a desideratum to have it sweetened by a pleasant syrup in proper proportion. We add, therefore, the words [take] "of syrup of almond"—quantity to be about one-quarter the whole bulk of the mixture. The remainder of the bulk is to be made up of water, but it lastly suggests itself to us to take, not plain water, but an aromatic water, in order still further to improve taste. We select *water of orange flowers*, and may write for this in one of three ways. We may, as in the case of the other ingredients, say [take] "of water of orange flowers" the requisite quantity, or we may say [take] "water of orange flowers up to the total measure of" the full bulk to be occupied by the mixture, or, again, "of water of orange flowers as much as may be necessary to attain" the same total bulk. Next, the necessary pharmacy in the case being fulfilled by the simple mixing of the ingredients, we append the word "mix," and follow with the direction "mark [it]:—'Teaspoonful thrice daily before eating.'" Then we date and sign the paper, and, if we please, order it "not to be renewed"; or, what is better, we use privately printed prescription-blanks of our own, having, if we so wish, "not to be renewed" printed at the top, and our *name*, *residence*, and *office-hours* printed at the bottom. So devised, such skeleton of prescription will look, in English, thus, where it will be observed that the due order of naming—first *basis*, and then, severally, *adjuvant*, *corrigent*, and *vehicle*—has been followed:

EXAMPLE I.—

Not to be Renewed.

For Mrs. A. B.

Take, Of Quinine Sulphate [quantity x],
Of Aromatic Sulphuric Acid [quantity y],
Of Syrup of Almond [quantity z],
Of Water of Orange Flowers [quantity w],
Or, Water of Orange Flowers up to [the measure of] [quantity n].
Or, Of Water of Orange Flowers as much as may be necessary to [attain the measure of] [quantity n].

Mix. Label—"Teaspoonful thrice daily before eating."

C. D., M.D.,
No. 1, First Street.

Office-Hours: 8 to 10 A.M., 5 to 6 P.M.
August 25, 1894.

To latinize, we find that the dictionary-words for *sulphate* and *quinine* are, respectively, "sulphas" and "quinina," and that our English model shows that both must be, as usual in prescription-form, in the *genitive.* Turning to our table showing guide to declensions, we find "ending in *-as*, genitive *-atis*, *acetas* (acetatis) and all salt names in *-as*." Evidently, then, *sulphas* belongs to this category, and its genitive, therefore, is *sulphatis.* As to *quinina*, the table declares all nouns in *-a*, with the exception of a brief list, to be of the first declension; so, turning to our other table, showing declensions, we find therefrom the genitive-ending *-æ* for the first declension, and thus make from *quinina*, *quininæ*. Accordingly we set down the first line of our example:

℞. *Quininæ Sulphatis*..... ....

The next entry, "aromatic sulphuric acid," offers a *noun*, "acid," with two modifying *adjectives*, "aromatic" and "sulphuric." Here, then, in the case of the word for "acid," we must have regard to *gender* as well as to declension, in order to know how to dress the adjectives. The dictionary gives for "acid" "*acidum*," and our table declares all nouns in *-um* to be of the *second declension*, and of *neuter gender*. The declension-table next affords *genitive* in *-i*, so that "of acid" becomes *acidi*. Turning now to the adjectives, we find that "aromatic" and "sulphuric" are, respectively, in dictionary-naming, in Latin, *aromaticus* and *sulphuricus*. To find the neuter thereof—since neuter the adjectives must be to agree with the neuter noun *acidum*—we look to our table of declension-schemes of adjectives, and find that adjectives of *-us*, in the masculine, make their neuter in *-um* of second declension. Our adjectives, then, are, in nominative, *aromaticum* and *sulphuricum;* but since they must follow their noun, not only in gender, but also in case and number, we must, in our prescription, turn these nominatives into genitives singular to obey the condition imposed by genitive singular noun *acidi*. So, then, *aromaticum* and *sulphuricum*, being forms in second declension, exactly as *acidum* itself happens to be, make, like *acidum*, genitive in *i*, and so become, respectively, *aromatici* and *sulphurici*. Then, as to order, adjectives in Latin follow their nouns, and the one of closest relationship takes precedence. The English order, then, suffers exact reversal, and the line must read:

Acidi Sulphurici Aromatici ....

Next is the entry, of "syrup of almond," a phrase presenting two nouns in the genitive. "Syrup" is *syrupus*, and our table shows nouns in *-us*, with a few exceptions, to be of the second declension. "Of syrup" then becomes *syrupi*. "Almond" is *amygdala*, which, ending in *-a*, like *quinina*, plainly is of the first declension, and makes genitive, therefore, *amygdalæ*. As to order, here we have a *pharmaceutical preparation* to deal with, in which case, as already pointed out, we cling to the English order, and let the dependent noun follow its leader. So we do not reverse, but set down:

Syrupi Amygdalæ........

Of the last entry, the first form is "of water of orange flowers," which, in Latin idiom, is expressed *of water of flowers of orange*—three nouns in the genitive, and one of them, "flowers," in the *plural* number. *Aqua,* "water," gives, as we may now divine without referring to the tables, *aquæ* for its genitive. "Flower" is *flos,* and the table of endings cites *flos* as being of the third declension, genitive *floris.* To find the genitive *plural,* we revert to the table of declensions, and find the ending to be, generally, *-um,* to be applied to the root of the word. The root of nouns of the third declension is found by subtracting the genitive singular case-ending, "is." Subtract, then, "is" from "floris," the genitive singular, and we have, for root or "stem," *flor-*, to which affix the genitive plural case-ending *-um*, and we have as the word we seek, *florum*, "of flowers." "Orange" is *aurantium*, which our experience with *acidum*, *aromaticum*, and *sulphuricum*, respectively, teaches us, at once, will make genitive *aurantii*. For the order of words, we note, first, that we have again a pharmaceutical preparation, so that *aquæ*, by right of might, comes first; but the pair of nouns "of flowers of orange" must behave with true latinity and range themselves "of orange [the] flowers," as, indeed, is in this case the common English idiom. The line then reads:

Aquæ Aurantii Florum.......

The second form of the same entry is "water of orange flowers up to [the measure of]." Here *water* becomes the immediate "object" of the verb *take*, instead of being, as before, a dependent upon the word for quantity. The case, then, must no longer be *genitive*, but *accusative*, and so from our declension-table we derive the accusative form *aquam*, which we must substitute for "aquæ." Next, the new form presents for rendering into Latin the phrase "up to [the measure of]." Turning to our table of odd words, we find the preposition *ad*, signifying "to," or "up to." The whole phrase then becomes:

Aquam Aurantii Florum, ad ......

wherein the words "aurantii" and "florum" suffer no change, for, of course, the phrase is still "water *of* flowers *of* orange."

The third form of the entry is "of water of orange flowers as much as may be necessary to [attain the measure of]." Evidently "of water" is, as at first, again *aquæ*, and the sole novelty of the present form is the introduction of the phrase "as much as may be necessary to," etc. For the translation of this our table of odd phrases may again be appealed to, and we find the set phrase *quantum sufficiat*, meaning "as much as may be necessary." Then the already found preposition *ad*, "up to," completes the phrase, and the line in its third form becomes:

Aquæ Aurantii Florum, quantum sufficiat ad..

But the phrase "quantum sufficiat," being a set phrase of common occurrence, is, as usual under such circumstances, *abbreviated*, and is expressed by the initial letters of the two words of its constitution, thus—*q. s.* The abbreviated reading of actual usage will then be:

Aquæ Aurantii Florum, q. s. ad....

Then the setting down of the symbols *M.* for "mix," and *S.* for "label," completes the translation.

Let us next illustrate other features of prescriptions by a different example. We propose for Mr. E. F. some castor-oil, and to make the vile stuff less disagreeable to the taste we will emulsify it, and then further dilute the emulsion with a sweetened and pleasantly flavored vehicle.

But for this end it will not do, as in the previous example, to order a simple admixture of all the ingredients, for the reason that emulsification is a peculiar process which requires that the oil and the emulsifier shall be rubbed together alone. Then when emulsification is accomplished, but not till then, may we add the flavoring and diluting ingredients. Here, then, we have distinct steps in the compounding, which steps *ought* to be properly detailed in the prescription. Such a prescription, then, would read, taking the yolk of an egg for the emulsifying agent, and a mixture of syrup of orange and spearmint water for the flavored diluent, as follows:

EXAMPLE II.—

For Mr. E. F.

Take, Of Castor-oil [quantity x],
The yolk of one egg:
Rub well together: then add
Of Syrup of Orange [quantity y],
Of Spearmint Water [quantity z].
Mix. Let it be made into an emulsion.
Label—"One-half at a dose."

G. H., M.D.,
No. 2, Second Street.

Office-Hours: 11 A.M. to 2 P.M.
August 26, 1894.

It is well in this case to add the words "let it be made into an emulsion," in order that the compounder may be certain to understand the aim of the prescription and so continue the rubbing of the oil and the egg-yolk until a good emulsion is formed. To translate now: "of castor-oil" is, structurally, "of oil of castor"—*oleum* is "oil," which we know enough now to turn at once into genitive *olei;* and for "castor" we have, in official nomenclature, the genus-name of the plant furnishing the oil, namely, *ricinus.* And *ricinus,* being, as is the case with the immense majority of nouns in *-us,* of the second declension, gives us, for "of castor," the genitive *ricini.* In order of words, the oil, being, practically, a *preparation,* the English order "oil of ricinus" obtains, and the entry appears:

℞. *Olei Ricini*.......

Next comes the unusual phrase "the yolk of one egg." Here, in the first place, the word *yolk* is plainly the direct object of the verb *take,* for the direction is not to take any particular quantity *of* yolk, but to take bodily, so to speak, *the* yolk of one egg. "Yolk" must then be in the accusative: *vitellus* is yolk, second declension, again, and declension-table gives accusative singular *vitellum.* Next, the phrase "of one egg" involves genitive of noun *egg*, with which must agree in gender, case, and number the numeral adjective *one*. *Ovum*, "egg," like *acidum* of former example, will make genitive *ovi*, and will require its adjective to assume the neuter gender. But *unus*, "one," is peculiar in declension, and, as shown in full on our declension-table, makes genitive for all genders in the irregular form *unius*. In order of words the practice in inditing preparations is here generally followed, so that *yolk* precedes *of egg*, though the adjective *one* follows the same. The translation is then:

Vitellum Ovi unius.

Next the pharmaceutical direction, "rub well together, then add," can be translated at once, by rote, from the table of odd words, thus:

Tere bene simul: dein adde

"Of syrup of orange," the phrase which next presents, we can render by former experience at once:

Syrupi Aurantii .. ....

"Of spearmint water" means "of water of spearmint." "Of water" we already know to be *aquæ*, and, for the official Latin name of *spearmint*, we have a translation of the words *green mint*. "Mint" is *mentha*, genitive, obviously, *menthæ;* "green" is *viridis*, an adjective, of course, whose genitive must be of the *gender*-form required by *mentha.* And the same *mentha*, being of first declension, is feminine. Turn we, then, to table of schemes of adjectives, and we learn that adjectives in *-is* have masculine and feminine in *-is*, which form is of the third declension with genitive in *-is.* *Viridis*, nominative masculine, is then also *viridis*, nominative feminine, and, further, gives *viridis* for the genitive of all genders. *Viridis* it is, then, and, adjective following noun, the whole reads:

Aquæ Menthæ Viridis........

Lastly, comes the further pharmaceutical direction, "*Mix:* let it be made into an emulsion." "Mix" becomes, as usual, *M.;* "let it be made into" is found by the table of odd words to be expressed by the single word *fiat*, which, being a verb in the passive voice, takes the *thing into which* the making is to be, as predicate-nominative. Hence *emulsum*, "emulsion," stands in the nominative as *emulsum;* the whole reading:

M.: fiat emulsum.

And, lastly, as always, "label" becomes *S.*

A third example: we design for Miss I. J. some potassium citrate. As is the general custom in the case of soluble salts whose dose is considerable, we will give this medicine in solution, and it is agreeable so to flavor the draught that the same shall taste like lemonade. Now, the easiest way to get potassium citrate in solution is to *form* the salt by addition of potassium *carbonate* to a solution of free citric acid. In such mixture, the stronger citric ejects the weaker carbonic acid of the carbonate, and, taking the potassium to itself, forms potassium citrate, which salt remains in solution. So our procedure in the premises will be this: we will order a solution of citric acid, flavor the same with a little of the essential oil of lemon, the better to imitate lemonade-flavor, and then (chemically) *saturate* the solution with potassium carbonate—that is, add gradually the latter salt so long as the evolution of bubbles of carbonic acid gas shall show that some citric acid still remains free, ready for attack upon more carbonate. Now, for the flavoring with the drop or two of oil of lemon, we must, as in the emulsifying of the castor-oil in the last example, observe certain precautions. If we add the lemon-oil to the *solution*, whether before or after the addition of the potassic salt, such oil will, because of its great insolubility in water, not become well diffused throughout; but if we rub it in a mortar with the dry crystals of citric acid, we shall, in the trituration, so break up the oil into minute globules that, when the impregnated acid comes to be dissolved, the oil will uniformly impregnate the solution. So, then, in this prescription, three distinct steps are required in the compounding, all of which should, of course, be described in our order.

The prescription, then, will read :

EXAMPLE III.—

For Miss I. J.

Take, Of Citric Acid [quantity *x*],
 Of Oil of Lemon [quantity *y*] ;
 Rub together ; then add
 Of Water [quantity *z*] ;
 Dissolve, and add, gradually,
 Of Potassium Carbonate as much as may be necessary up to saturation.
Label—" A teaspoonful as occasion arises."

K. L., M.D.,
No. 3, Third Street.

Office-Hours : 9 to 11 A.M.
August 27, 1894.

To latinize : *acidum*, " acid," we have had before ; *citricus*, " citric," will, like *sulphuricus* of our former example, make neuter *citricum*, to agree with *acidum*, and then for the genitive, to signify "of citric acid," the form will be :

℞. *Acidi Citrici*........

Next, *oleum*, " oil," we have also had before ; *limon*, " lemon," requires reference to the table of endings, where we find the word cited as belonging to the third declension, genitive *limonis*. " Of oil of lemon " then becomes :

Olei Limonis........

" Rub together ; then add " has also occurred before, as thus :

Tere simul : dein adde.

" Of water," also of former occurrence, is *aquæ* from nominative *aqua :*

Aquæ........

Next, " dissolve " is found by the table of odd words to be *solve;* " and " to be *et*, " add," *adde*, and " gradually," *gradatim*. Thus this next line reads :

Solve, et adde, gradatim.

" Of potassium carbonate " presents next. " Potassium " is *potassium*—all titles of metals in -*um* being embodiments into English of a Latin word. " Of potassium " is, then, *potassii*, and "carbonate" being *carbonas*, a word belonging to the large group of salt-names in *-as*, makes genitive, analogous to the genitive of *sulphas*, already considered, *carbonatis.* "Of potassium [the] carbonate" then reads:

Potassii Carbonatis........

Lastly comes the phrase, "as much as may be necessary up to saturation," which, by our vocabulary of odd words, we render,

Quantum sufficiat ad saturandum,

or, as before, abbreviating *quantum sufficiat*, the full line reads:

Potassii Carbonatis q. s. ad saturandum.

A fourth example will illustrate additional points. We wish to give Mr. M. N. a short course of laxative pills, to be composed of equal parts of "blue pill," aloes, and rhubarb. We therefore write for the proper quantity of these several articles, and direct the pharmacist to mix the materials, adding thereto what *water* may be necessary to give proper cohesiveness to the mass, and then to divide the same into a specified number of pills. Lastly, the box is to be labelled, "Two pills, at bedtime." Such prescription will read:

EXAMPLE IV.—

For Mr. M. N.

Take, Of Mass of Mercury [quantity x],
Of [powder of] Purified Aloes [quantity x],
Of [powder of] Rhubarb [quantity x],
Of Water, as much as may be required.
Mix, and divide into n pills.
Or, Mix: let it be made into a mass to be divided into n pills.
Label—"Two pills, at bedtime."

O. P., M.D.,
No. 4, Fourth Street.

Office-Hours: 8 to 9 A.M., 4 to 5 P.M.
August 28, 1894.

In this example, the expression "powder of" is interpolated in brackets before the official title of the drug, the point being this: *Purified aloes* is a stuff in lumps, and *rhubarb* is a root in bulk, and in order to embody such matters into a pill-mass they must first undergo pulverization. Such pulverized article the pharmacist will certainly take, whether the prescription order the form of powder or not, and thus both practices obtain among prescribers—some consider it enough for the physician simply to designate the drug he wants by its pharmacopœial title, leaving it to the pharmacist's knowledge of his own business to prepare such drug properly for the fulfilment of the prescription ; while others think it safer to state in the prescription that the drug is to be taken in powder. Another point is that in ordering the *water* required, it is wholly unnecessary to write out "as much as may be required *to bring the mass to a proper pilular consistence*," for the pharmaceutical purpose of the water is obvious, and the compounder may therefore safely be left to divine the same. Indeed, many prescribers would leave out all mention of the water, holding it to be enough for the physician to state that he wants such and such drugs made into a pill-mass; what "excipients" pharmacy may require for the purpose being held to be a thing of which the pharmacist is the best judge. A third point in this prescription is this: It will be recalled that the pills were to be made of equal parts of the three ingredients, and the example, therefore, reads that, of each of the three, the same quantity, "x," is to be taken. Now, wherever two or more drugs are to be taken in identical quantity, the practice is, for shortness' sake, to write in this form:

Take, Of A } of each [quantity x].
Of B }

Hence our present prescription would, in actual practice, be begun thus:

Of Mass of Mercury, Of [powder of] Purified Aloes, Of [powder of] Rhubarb,	of each [quantity x].

To translate now into Latin: "mass" is *massa*, and "mercury" is *hydrargyrum*, and our frequent experience of the endings hereby presented enables us to write genitives at once:

Massæ Hydrargyri......

wherein *massæ* precedes *hydrargyri* because the phrase is the title of a preparation. In the next entry, "powder" is *pulvis*; nouns in *-is*, by reference to table of endings, are of the third declension, and of very diverse genitives. We search carefully, and at last find catalogued "*pulvis* (*pulveris*)." *Pulveris* is, then, the genitive we seek, and, by the way, a safe guide to these troublesome third-declension genitives is afforded in very many cases by English words derived from the Latin, wherein the *root* of the word is displayed. Thus, in the present instance, the words *pulverize* and *pulverulent* furnish the key, showing at a glance the root *pulver-*, whence, of course, genitive *pulveris*. "Of purified aloes" will stand, in Latin order, the adjective following its noun—"of aloes purified." "Aloes" is *aloë*, the dieresis "¨" being put over the *e* to show that such letter forms a syllable by itself and is not part of a diphthong *œ*. Of this word *aloë*, now, we must find genitive and *gender*, the latter because of there being an adjective to be properly conformed to the noun. Our table of endings shows *prescription*-nouns in -*e* to be certain of Greek origin assigned to the first declension, and, therefore, of the feminine gender. But in inflection, in the singular number, these nouns are peculiar, and our declension-table must, therefore, be referred to; this is done, and the genitive is found *aloës*. Then "purified" is *purificatus*, which, in feminine form, is *purificata*, of first declension, genitive therefore *purificatæ*. The whole title, then, becomes:

[*Pulveris*] *Aloës Purificatæ*.......

Next, "rhubarb" is *rheum*, and at once we arrive at the rendering:

[*Pulveris*] *Rhei*........

Then, "of each," we find, by reference to list of odd words, to be the Greek *ana*, embodied into prescription-Latin. And this word, being of frequent occurrence, is commonly abbreviated and takes form āā. Our triple basis is then written:

Massæ Hydrargyri,
[Pulveris] Aloës Purificatæ,
[Pulveris] Rhei, āā [quantity x].

"Of water, as much as may be required," we write, without ado,

Aquæ, quantum sufficiat,

or, abbreviated,

Aquæ, q. s.

Of the two forms of writing the pharmaceutical direction in the case, the first, "mix, and divide into n pills," is phrased, in Latin idiom, "mix, and divide into pills to the number of n." The table of odd words gives *M.* for "mix," *et* for "and," *divide* for "divide," and *in* for "into," to be followed by noun in the *accusative*. *Pilula* is "pill," and, being of first declension, gives accusative plural, "pills," *pilulas.* "To the number of," is found among odd phrases as *numero*, which is again an oft-recurring word and so suffers abbreviation, being commonly written *no.* The direction, then, reads:

M., et divide in pilulas no. [n].

The other form, "Mix : let it be made into a mass to be divided in n pills," would be in Latin style, "Mix : let [it] be made [into] a mass, into pills to the number of n to be divided." Here, then, we introduce, from our vocabulary of odd words, *fiat*, "let [it] be made," and *dividendus*, "to be divided." And "mass," as the table directs, will then be in the nominative, predicate, and *dividendus*, being an adjective, will be required to agree with the word for "mass," in gender, number, and case. Hence *massa*, first declension, feminine, nominative singular, will require *dividendus* to become *dividenda*, and the line will read:

M.: fiat massa in pilulas no. [*n*] *dividenda.*

Thus, in full, the two forms of phrasing the pharmaceutical direction in question; but, obviously, such forms will be matters of staple occurrence, being used whenever the substance—or admixture of substances—prescribed is to be *divided* by the pharmacist, whether into pills, packages ("powders"), capsulefuls, troches, suppositories, or other specialized forms. Hence—the old story—*abbreviation* is the fashion, and the above lines will more commonly be written, in actual prescribing, as follows:

M., et div. in pil. no. [*n*]

and,

M.: ft. mass. in pil. no. [*n*] *dividend., or div.*

Similarly, the phrases, *in chartulas*, "into packages of powder"; *in capsulas*, "into capsulefuls";

in trochiscos, "into troches"; *in suppositoria*, "into suppositories," become, severally, *in chart.*, *in caps.*, *in trochisc.* or *in troch.*, and *in suppos.* Such abbreviations, however, are distinctly *not* to be recommended, for abbreviation leads to error "as the sparks fly upward."

A fifth example will illustrate yet another point. We want for Miss Q. R. a mild chalybeate, and we find provided by the Pharmacopœia certain appropriate pills, the so-called "pills of ferrous carbonate," where, furthermore, the standard weight of each pill is handy for the present indications of dosage, two of the pharmacopœial pills being just the proper amount for a single administration. Our prescription, then, in this case, is simply the form of an order to the pharmacist to dispense so many of these pills, and label the box "two pills, thrice daily." Thus:

EXAMPLE V.—

For Miss Q. R.

Take, Pills of Ferrous Carbonate to the number of [*n*].

Label—"Two pills, thrice daily."

S. T., M.D.,
No. 5, Fifth street.

Office-Hours: 9 to 1,
August 29, 1894.

In latinizing, we observe that the word *pills* is, in this prescription, the immediate object of the verb *take*. Here, then, is one of the exceptional instances where the title of the medicine, in prescription, stands in the *accusative* instead of in the genitive. And the accusative plural of *pilula*, "pill," we have already found to be *pilulas*. *Ferrous Carbonate* is, in the Latin of the Pharmacopœia, the translation of the simple title *Iron Carbonate* (meaning carbonate *of* iron). "Iron" is *ferrum*, with genitive *ferri*, and "carbonate" we have already found to be *carbonas*, with genitive *carbonatis*. The prescription will therefore read:

℞. *Pilulas Ferri Carbonatis, no.* ⌊*n*].
S.—"Two pills, thrice daily."

Suppose, next, that for a half-grown girl we want this same chalybeate preparation, but need now, for each dose, a little more than one pill, but less than two, of those of pharmacopœial weight. Nothing is simpler; we estimate our quantities for the getting of a pill of somewhat less weight than the standard, and order the pharmacist, accordingly, to take such prescribed quantity of "pills of ferrous carbonate" and divide the mass into so many pills. We write, that is—

EXAMPLE VI.—

For Miss U. V.

Take, Of Pills of Ferrous Carbonate [quantity *x*].

To be divided into [*n*] pills.

Label—"Two pills, thrice daily."

W. X., M.D.,
No. 6, Sixth Street.

Office-Hours: 3 to 5 P.M.
August 29, 1894.

Here, evidently, "pills" with its adjective reverts to the genitive, and so the title of the drug appears:

℞. *Pilularum Ferri Carbonatis* [quantity *x*].

"To be divided into [*n*] pills" is a phrase of now familiar structure, except that in this instance the *thing* "to be divided" is, grammatically, *quantity x*. The gender, number, and case, therefore, of *dividendus* must here conform to those elements of the word expressing denomination of quantity. Such word will, of course, be in the *accusative case*, but its gender and number will depend on circumstances. "Grains" or "grammes" will be, respectively, *grana* and *grammaria*, accusative plural, neuter, of second declension, and the adjective will then be *dividenda*. "Scruple" gives *scrupulum*, and "scruples" *scrupulos;* "drachm" and "drachms," "ounce" and "ounces," give, respectively, *drachmam*, *drachmas*, *unciam*, and *uncias*. And, to conform, we shall thus have, severally, of the adjective, *dividendum*, *dividendos*, *dividendam*, *dividendas*. Our line will therefore read :

In pilulas no. [*n*] *dividenda*, or *-um*, or *-os*, or *-am*, or *-as*.

A few new points are exemplified in a seventh instance. We want, for personal surgical use, a solution of lunar caustic of a certain strength. The prescription, in such case, takes the simple form of ordering the proper proportion of the caustic and of *distilled* water (which alone should be used for the purpose), and directing, then, that the caustic be dissolved in the water. In this instance, the medicine being for our own use, a *labelling* is, from one point of view, supererogatory ; but yet, having due regard to the freaks of the imp of misadventure, it is best to take every precaution, and so direct a label to read, " for external use." Our prescription will, then, be :

EXAMPLE VII.—

For Self.

Take, Of Silver Nitrate [quantity *x*],
Of Distilled Water [quantity *y*].
Dissolve, and label: "For external use."

Y. Z., M.D.,
No. 7, Seventh Street.

Office-Hours: 2 to 4 P.M.
August 29, 1894.

Latinized, *argentum*, "silver," and *nitras*, "nitrate," give, obviously, the reading in the genitive:

℞. *Argenti Nitratis.........*

"Of distilled water" is, in Latin order, "of water distilled," and is rendered

Aquæ Destillatæ........

And "dissolve and label" appears as

Solve et S.

in which phrase observe that the familiar word, "mix," does not occur, for here is no *mixing* proper, but a simple *dissolving*.

So we might multiply examples, but enough have now been given to illustrate the commoner run of prescription-forms, and to show how easy of application are our few rules and tables of Latin words and usages.

III. *The Computing of Amounts in Prescriptions.*—We pass from the subject of form and language to that of the art of *computing amounts* in prescribing. This matter has, so far, been purposely passed by in our discussion and exemplification of prescription-technics, because, being an entirely independent consideration, it is best studied by itself.

The subject of amounts, in prescription-writing, divides into two parts: first, the consideration of the total amount of the mixture, and, secondly, of the relative proportion of the ingredients. As regards totals, the rule obtains, at the outset, *not to order more than the present prognosis seems to call for.* To prescribe two dozen pills when half a dozen only prove to be needed, or a four-ounce mixture of which but a few teaspoonfuls are taken, argues—so the patient naturally reasons—either carelessness or ignorance on the part of the prescriber. It is better, therefore, to order too little than too much, letting the prescription be renewed if the first quantity prove insufficient. Various prudential considerations also argue against a procedure that allows half-used parcels of medicines to "lie around loose" about the household. The very first step, then, in determining amount, is to think, and think *carefully*, about how much of the medicine seems to be needed. Such estimate, if the medicine be for external use, as in the case of an ointment, a liniment, or a wash, must be based on general considerations of how and how often the thing is to be applied; but if the medicine is to be taken internally, the estimate is figured out from the *number of doses*, first, and *dimensions* of dose, secondly. We say to ourselves that x doses of the basis will probably do the work, and, then, that each dose shall, in the mixture as actually administered, occupy a dimension y. The total bulk, therefore, of the mixture is x times y dimension, or x times y weight, as the case may be. Having thus arrived at *about* the total required, for the *exact* amount we select a quantity representing a convenient round number in terms of the system of weight or measure by which we are prescribing: this for the reason that the arithmetic of apportioning the constituent parts of the mixture is thus rendered simpler than would be the case were an odd quantity taken.

In the case of fluids, furthermore, we select round numbers for the additional reason that medicine-vials are made of stated capacities corresponding to evenly expressed dimensions in apothecaries' and metric measure, respectively, and our prescribed mixture ought, both for the sake of convenience and of elegance, to measure just a bottleful.

Now, these same convenient round numbers that we pick upon to determine our totals by, *will differ according to the genius of the system of weight or measure* that we employ. Thus, in the apothecaries' system, the relation of the denominations is—disregarding the comparatively little-used scruple—essentially on a *duodecimal* basis. In using this system, therefore, we naturally calculate by numbers bearing a simple ratio to the number 12, viz., 2, 4, 6, 8, 12, 18, 24, 60, 120, 180, 240, 480; and hence medicine-vials, made to accord with apothecaries' measure, are made of the several capacities *one*, *two*, and *four fluidrachms*, and *one*, *two*, *four*, *six*, *eight*, and *twelve fluid-ounces*. In using the *metric* system, however, the most essential feature of which is the *decimal* relation of its denominations, it is most natural, and is infinitely easier, to deal with values whose expression is by those numbers simply related to the number 100, viz., the several numbers 2, 5, 10, 20, 25, 50, 75, 100, 200, 250, 500, 1,000. This point is often overlooked by novices in the use of the metric system, and the stupidity is committed of *estimating in duodecimals*, and then *expressing in decimals*—a procedure as clumsily absurd as would be the calculation in pounds, shillings, and pence of prices which are to be stated in terms of dollars and cents.

Having settled upon the total, the apportioning of the constituents proceeds thus: The amount of the *basis* will be as many times the quantity intended for each dose as there are doses in the total mixture; the amount of the *adjuvant* or *corrigent* will generally be determined by that of the basis, and the ingredients that go to form the *vehicle* will, then, fill the measure or weight of what remains of the total after allowing for the measure or weight already taken by basis and adjuvant, *except* in the case of prescribed weights of solids to be dissolved in volumes of fluid. In such case, we have to remember the peculiar physical fact that a solid dissolved in a fluid does not increase such fluid's bulk by the full measure of its own volume—in fact, increases it so little that, in the generally feeble solutions ordered in medical prescribing, it is customary to disregard altogether the volume of a dissolved solid, and to compute the volume of the solvent by the full capacity of the bottle intended to be filled.

Such are the principles for computing amounts in prescribing; and now, as in the dealing with form and language, it is wise to illustrate by examples. We will, then, take the examples already presented, and proceed to fix the amounts of the ingredients by the apothecaries' and metric systems, severally.

The first prescription was for stomachic doses of quinine sulphate, to be dissolved, by the aid of aromatic sulphuric acid, in a mixture of syrup of almond and water of orange flowers. This being a medicine intended for internal use, the first step toward determining total measure or weight is to bethink us how many doses are likely to be needed. And by the outlook of this case we will assume the medication probably required to be three doses daily for three or four days—that is, then, we need to prescribe an aggregate of somewhere between nine and twelve doses. Then, as to *dimension* of dose, which comes next to be determined ; inasmuch as the quantity of the quinine salt in each dose is to be small, the medicine being for a stomachic effect only, a *teaspoonful* of fluid will be a convenient measure to hold the same. Furthermore, since the medicine is innocent, and the indications for dosage not very exact, it will, in this case, be perfectly legitimate to allow the doses actually to be dispensed by that inaccurate measure, a common teaspoon. We thus gain the preliminary point—independent, be it observed, of the particular system of weight or measure by which we propose to prescribe—of proposing an aggregate of somewhere between nine and twelve teaspoonfuls of mixture. Now, however, the determination of the *exact* total of the mixture, and the apportioning of this among the several constituents thereof, will be affected by the scale of weight or measure adopted. We will proceed first by the apothecaries' system. An aggregate of somewhere between nine and twelve teaspoonfuls will suggest what round number of apothecaries' measure denominations? Instantly we say, *two fluidounces* average *twelve teaspoonfuls* of the rather large teaspoons of the present day (six teaspoons to the fluidounce); hence let our aggregate be two fluidounces, and let us figure on a basis of twelve teaspoonful doses. This point being settled, we are ready for the apportioning, and in doing this it will be found handiest to *write down first* the titles of *all* the constituents, and then, thinking of nothing else, compute and set down the amounts for each. The computation in this case, then, proceeds thus: Of the quinine salt we want *about* one grain for each dose; let us assume the dose, then, at that convenient round number, *one grain* exactly; then twelve one-grain doses give, at once, *twelve grains* for the total of the basis in the mixture. We accordingly set down:

℞. *Quininæ Sulphatis* gr. xij.,

in which form observe that *gr.* and *not* "*grs.*" is the abbreviation of the Latin for "grains"; also that the number *twelve* is written in *Roman* numerals *following* the symbol for denomination.

Next, as to the amount of the thing that is to help the basis—in this case, the acid that is to determine the solution of the quinine salt. The effect is here a chemical one, and the amount of the acid is, therefore, determined by the amount of the quinine. In a rough way—accurate enough for prescription-needs—it may be accepted that to effect solution of a given quantity of quinine sulphate, one-and-a-half times such quantity of aromatic sulphuric acid will be required. And this ratio, it is also accurate enough for us, may be estimated in terms of apothecaries' *measure*. Hence, having *twelve* grains of quinine sulphate, we shall need *eighteen* of the analogous denomination of volume of the acid—in short, eighteen *minims*. We write, then :

Acidi Sulphurici Aromatici ♏ xviij

Coming now to the vehicle, we find this to be an admixture of two ingredients, and to know how to apportion these we must first note what measure the aggregate is to fill. We say *measure*, because in using the apothecaries' system it is customary to measure fluids. The aggregate of the vehicle will be so much as is left of the two fluidounces measure after putting in the quinine and the acid.

But here, on the principle already expounded, we entirely disregard the minute effect on volume exercised by the presence of the quinine salt in solution, and even the space occupied by the eighteen minims of acid is hardly worth considering. We practically estimate, then, that we have the total measure of two fluidounces to be filled by the vehicle, and, a fair proportion between our ingredients being one part of syrup to three of water, we apportion the two fluidounces thus:

Syrupi Amygdalæ...........f ℥ ss.
Aquæ Aurantii Florum......f ℥ iss.

Here, in the matter of form, note the abbreviation *ss.* standing for *half*, derived from the Latin word *semis*, "half"; and note in the expression for denomination that the letter *f* precedes the symbol "℥," signifying that it is *fluid* ounce, a measure of capacity, and not *ounce*, a weight, that is signified. This same letter, *f*, which should thus properly always be placed before the symbol when expressing fluidrachms or fluidounces, is often omitted, on the idea of taking it for granted that in dealing with fluids the pharmacist will measure and not weigh. But accuracy, like honesty, is always the best policy, and this omission is, therefore, not to be recommended.

To return to the prescription : since the vehicle is thus prescribed to the full measure of two fluidounces, if the pharmacist mixes all the ingredients in a graduate, and then pours into a two-ounce vial for dispensing, there will be a surplus of eighteen minims of the mixture, by reason of this measure of acid disregarded in allotting the measure of the vehicle ; and there will be a slight error in the proportion of quinine in the mixture, because the total is eighteen minims in excess of the amount originally assumed. Both these errors are trivial, but they can, if thought better, perfectly be avoided by the following simple procedure: Let us, of our last and least important ingredient, the water, not order the *fixed amount, one-and-a-half fluidounces*, which makes the whole eighteen minims too much, but direct that water be " taken " *until the whole mixture shall be brought to the final full measure of two fluidounces*. Now, then, the pharmacist, having put into his graduate, or into the two-ounce dispensing-vial, the three first-named ingredients, simply fills up to the two-ounce mark on the graduate, or to the neck of the bottle, with the orange-flower water. This procedure is evidently to be commended, for the reason that it is more accurate, on the one hand, and handier in its working on the other, both to prescriber and compounder. Following such method, the last entry will, then, appear thus:

Aquam Aurantii Florum ......ad f ℥ ij.
Or, *Aquæ Aurantii Florum*...q. s. ad f ℥ ij.

Of these two forms, both of which mean precisely the same thing, the latter is perhaps the more advisable, since the introduction of the abbreviation *q. s.* calls more especial attention to the nature of the order. In both forms, the *amount*, be it observed, now appears as *two*, and not, as before, *one-and-a-half* fluidounces, although it is practically even less than one-and-a-half fluidounces that is taken; but the word "two" is but part of the phrase, "up to the full measure of two," signified by the prefix of the preposition *ad*, as already sufficiently explained.

Next, to determine the amounts of the ingredients for the same mixture by the *metric* system: here we do *not* fly to our table of equivalents between apothecaries' and metric sums, and translate into grammes our twelve grains, eighteen minims, half ounce, etc. We *may*, of course, do this, and many would-be learners of metric prescribing follow such method, thinking it the proper course; but such procedure, as already expounded, is to defeat the very advantage which the metric system offers—namely, the convenience of computing by *decimal* ratios. What we do, if we do wisely in the matter, is to go back to the beginning, and, with no more thought of grains and of ounces than if such things were not, think and figure solely in the terms and by the spirit of the metric system. In the present case, then, we return to the starting-point, that we propose somewhere between nine and twelve teaspoonful doses of a mixture containing in each teaspoonful a small charge of quinine sulphate. What, then, is the nearest round-numbered volume in metric denomination to the measure of between nine and twelve teaspoonfuls? Modern teaspoons average the capacity of *five* cubic centimeters, and so *fifty* cubic centimeters will be just the measure of *ten* teaspoonfuls. We just as naturally, therefore, now pitch on a decimal aggregate of *ten* doses, as before, when working a duodecimal system, we selected a total of *twelve*. And this aggregate of ten doses is the measure, *fifty cubic centimeters*—the measure of fifty grammes of water. To apportion the ingredients, we first, as before, bethink us what is to be the individual dose of the basis—the quinine. And the same is, for stomachic purpose, *about* five centigrammes (0.05 Gm.), and so *exactly* five centigrammes shall it be. But here some one will be certain to say: "Stop! you are not giving us, as you purport to do, the same mixture as before! Before, you assumed the patient to need *one grain*, per dose, of quinine, and one grain is the equivalent of *six-and-a-half* centigrammes, instead of the *five* that you are now proposing!" Perfectly true, but the point is this: When we declare a sluggish stomach to be in such plight that a gentle bitter-tickling will probably benefit it, have we any means of gauging with scientific precision *exactly* the proper size of the titillation? The initiated know well enough that we have not, so that when we solemnly order a *one-grain* tickle, they know that we do so, not because by any abstruse pathological observation we have discovered that one grain happens to be exactly the amount therapeutically indicated, but simply because that same one grain is *the nearest round-numbered quantity*, in *terms of apothecaries' weight*, to the average quantity required in the run of cases assumed. The same principle of *convenience*, therefore, which makes a prescriber by the apothecaries' system estimate a dose at exactly the round weight, *one grain*, leads him who uses the metric to assume the dose, in centigrammes, at the decimally handy figure *five*, and not *six*. Our total quantity of quinine sulphate is, therefore, for ten doses, ten times five centigrammes, *i.e.*, fifty centigrammes (0.05 Gm. × 10 = 0.50 Gm.). Hence we write;

℞. *Quininæ Sulphatis* 0.50 Gm.,

in which form note the *Arabic* numerals, and the position of the expression for gramme, Gm. *following* the numeral. Observe also the zero figure occupying the unit place of the integer, a point in notation that should never be omitted. For thus we assure the reader of the prescription that the decimal point—that point of weighty moment—*is* just where it is *intended* to be. Without such zero, that error of tenfold consequence, a misplacement of the figures in relation to the decimal point, is easy to commit, and, committed, would only be detected by the reader through a knowledge of dosage.

Next, as to the amount of acid, this, as previously shown, is to be half as much again as that of the quinine. Hence fifty centigrammes of quinine will require seventy-five of acid:

Acidi Sulphurici Aromatici . . 0.75 Gm.

Lastly, as to the vehicle, we will naturally, as before, disregard the volume occupied by the quinine in solution and by the acid, and hence consider that we have the whole of the fifty cubic centimeter bulk of our mixture to be apportioned between the two ingredients of the vehicle, the syrup and the water. Desiring, furthermore, *about* the proportion one to two or three, we shall very naturally assign *fifteen* cubic centimeters to the syrup and *thirty-five* to the water. But in order to get these volumes, we write—following custom in prescription-use of the metric system—for the proper *weights* in grammes, the pharmacist compounding by counterpoising his dispensing-vial on the balance and then weighing into it, so to speak, the several ingredients. What, then, is the weight of fifteen cubic centimeters of syrup? Fifteen cubic centimeters of *water* weigh fifteen grammes. But *syrup* is one of the few pharmaceutical bodies whose specific gravity is so different from that of water that we must take the same into account. And, by the table given awhile ago, we find that, pharmacopœial syrups being one-third again as heavy as water, we must order one-third additional in terms of weight to get a given volume. Wanting, then, fifteen units of volume, we order in units of weight fifteen plus one-third of fifteen—that is, *twenty*—and so set down:

Syrupi Amygdalæ 20.00 Gm.

As to the thirty-five cubic centimeters of orange-flower water, that is instantly disposed of: being an aqueous fluid, its weight is gramme for cubic centimeter, and so the thirty-five of measure is thirty-five of weight also. We order, then,

Aquæ Aurantii Florum . . . . . 35.00 Gm.

If now it be deemed advisable to take into account the bulk of the *acid*, as we did, when using the apothecaries' system, by the plan of writing for the orange-flower water to be added "*up to the full measure of*," etc., we can accomplish the end, metrically, with perfect ease, by deducting from the amount of water to be ordered the amount of acid already taken. Now, this was inconvenient in using the apothecaries' system, simply because of the want of relation, in that system, between denomination-values and ordinary arithmetical notation, making the calculation and the expression of odd amounts so clumsy as to be impracticable. That is, to estimate and write for the quantity *one-and-a-half fluidounces less eighteen minims* would be intolerably awkward ; but, in the metric system, the accord of the same with standard decimal notation makes such procedure perfectly simple. We have merely, as it were, to deduct seventy-five cents from thirty-five dollars—a sum done in the head on the instant—and the remainder, thirty-four dollars and a quarter, is as easy of expression as the original amount. To be exact, then, we give place in our estimate for the 0.75 Gm. of acid, and instead of ordering of the water 35.00 Gm., we write for 34.25 Gm. only.

Our first example, then, as it would actually be written in practice, will be, in full, as follows:

EXAMPLE I.—

Not to be renewed.

For Mrs. A. B

[*Apothecaries' System:* 12 *teaspoonfuls; dose of basis, gr. j.*]

℞. Quininæ Sulphatis............. gr. xij.
Acidi Sulphurici Aromatici. ... ♏ xviij.
Syrupi Amygdalæ............... f ℥ ss.
Aquæ Aurantii Florum............ f ℥ jss.
(*Or*, Aquam Aurantii Florum. .... ad f ℥ ij.)
(*Or*, Aquæ Aurantii Florum...q. s. ad f ℥ ij.)

[*Metric System:* 10 *teaspoonfuls; dose of basis,* 0.05 Gm.]

	Gm.
℞. Quininæ Sulphatis...........	0|50
Acidi Sulphurici Aromatici...	0|75
Syrupi Amygdalæ	20|00
Aquæ Aurantii Florum.......	35|00, *or*, 34|25

M S.—"Teaspoonful thrice daily before eating."

C. D., M.D.,
No. 1, First Street.

Office-Hours: 8 to 10 A.M., 5 to 6 P.M.
August 25, 1894.

Proceeding to our second example—castor-oil emulsified by the yolk of an egg, and the emulsion diluted with a mixture of syrup of orange and spearmint water—we arrive at our amounts as follows: This mixture, being a purge, is required in but single dose, and the total bulk hinges directly, then, on the dimension of the single dose. Now, Mr. E. F. is a big, hard-working man, and will probably take a full measure of the oil; so we will make our mixture to accord with a full dose of castor-oil, although, for safety's sake, we will order only one-half to be taken at once, reserving the other half for a possible repetition. Now, a full dose for a hearty man is, of this oil, from a tablespoonful to a tablespoonful and a half, the dosage having considerable range. To emulsify, we shall need, of egg-yolk, *about* one-half the amount of the oil, and, for a pleasant further dilution, shall want of the diluent between one and two measures of the emulsion. In all the amounts, from the nature of the case, there is a good deal of latitude. Starting with the amount of the oil, and computing first in the apothecaries' system, we shall naturally fix upon *one fluidounce* as being the nearest even measure available to represent a quantity of a fluid which shall be somewhere between one and one-and-a-half tablespoonfuls. Then half that amount of yolk of egg will be half a fluidounce, which happens to be just about the measure of an average single yolk. Having thus a fluidounce and a half, now if we make, of the whole, a four-ounce mixture, we shall have an exact bottleful, as bottles are made, and the two-and-a-half fluidounces to be occupied by the diluent will be within the range of advisable proportion of the same. So we order of the essential ingredients,

℞. *Olei Ricini*........................f ℥ j.
Vitellum Ovi unius.

Of the two-and-a-half fluidounces to be occupied by the diluent, we will assign the odd half to the syrup and the two to the water, taking but a small proportion of syrup, because of the viscidity already present in the oil-emulsion. So we write for the diluent,

Syrupi Aurantii..................f ℥ ss.
Aquæ Menthæ Viridis.............f ℥ ij.

Or, if we are pursued by a demon of exactitude, remembering that the yolk of egg may not turn out precisely a half-fluidounce, we save ourselves as to total bulk by writing the last entry,

Aquæ Menthæ Viridis....q. s. ad f ℥ iv.

Metrically, our range of quantity for the oil, to

be somewhere between one and one-and-a-half tablespoonfuls, will suggest the decimally convenient measure of *twenty-five cubic centimeters* (twenty cubic centimeters equalling the capacity of the average modern tablespoon), and the yolk of one egg in its entirety being handy to take, we will allow the same, although somewhat more than half the measure of oil. Estimating such yolk at fifteen cubic centimeters, this quantity, with the oil, gives a total bulk, for the emulsion, of forty cubic centimeters. And, at once, for the amount of diluent, the quantity *sixty* cubic centimeters suggests itself, a quantity which will bring the whole to the even measure of one hundred cubic centimeters; and this measure of sixty we will apportion by giving fifteen to the syrup and forty-five to the water—an apportionment that gives us a proportion between the amounts, respectively, of syrup and water which is in the neighborhood of that obtained in the prescription by apothecaries' measure. To order these several volumes, now, in terms of weight, we have to remember that the first substance, the oil, is one-tenth lighter than water, but yet, since the dosage, in the case of castor-oil, is somewhat indeterminate, the error of disregarding the difference of specific gravity will be of no practical moment whatever. Still, if we prefer, the needed allowance can be made without the slightest difficulty: twenty-five less one-tenth is twenty-five less two and a half, or twenty-two and a half. We order, therefore, for the emulsion,

℞. *Olei Ricini* ....22.50 Gm.
Vitellum Ovi unius.

As regards the diluent, the fifteen cubic centimeters of syrup will, as in the other example, weigh twenty grammes, and the forty-five of water, forty-five. The completed example then reads thus:

EXAMPLE II.—

For Mr. E. F.

[*Apothecaries' System: four-ounce mixture.*]

℞. Olei Ricini........................f ℥ j.
Vitellum Ovi unius.
Tere bene simul; dein adde
Syrupi Aurantii... .... f ℥ ss.
Aquæ Menthæ Viridis.......... f ℥ ij.
(Or, Aquæ Menthæ Viridis..q. s. ad f ℥ iv.)

[*Metric System: 100 C.c. Mixture.*]

	Gm.
℞. Olei Ricini...........................	22 50
Vitellum Ovi unius.	
Tere bene simul; dein adde	
Syrupi Aurantii..	20 00
Aquæ Menthæ Viridis...............	45 00

M. Fiat emulsum.

S.—" One-half at a dose."

G. H., M.D.,
No. 2, Second Street.

Office-Hours: 11 to 2.
August 26, 1894.

The third of our examples was for lemon-flavored citric acid, to be dissolved in water to the strength of ordinary lemon-juice, and then neutralized with potassium carbonate. Here our data for computing amounts are that we want about *five or six doses;* these of *tablespoonful* dimensions; that a proper strength of the potion is afforded by operating upon a *six-per-cent. strength of acid solution*—which is about the acid strength of average lemon-juice; and that about a *one-tenth of one per cent. impregnation with oil of lemon* gives the desirable degree of lemon-flavor to the solution. By the apothecaries' system we have *four fluidounces* as the average measure of six tablespoonfuls, and so we fix the first item, the totality of the prescription, at that figure. The next step is to find six per cent. of four ounces, which will be the amount of acid required—a clumsy problem in the apothecaries' system, but which we solve thus: *one grain* is the *one-four-hundred-and-eightieth* of one ounce; suppose it were the *one-five-hundredth* and it would be just the *one-fifth of one per cent.* of an ounce. Then *five grains* would be *one* per cent. of an ounce, and six times five, or *thirty*, would be *six* per cent. Then, further, if *thirty* grains be *six* per cent. of *one* ounce, *four times thirty*—one hundred and twenty—is six per cent. of *four* ounces. Such number, then, being the conveniently rounded amount of *two drachms*, we adopt as being near enough to what is demanded. The same arithmetic also fixes the one-tenth of one per cent. of oil of lemon at *two minims* (one per cent. of four fluidounces is, roughly, *twenty minims*, one-tenth of which is *two*).

Turn now from this roundabout process of calculation to the decimal convenience of the metric system. Needing five or six tablespoonfuls, we instantly select a total measure of one hundred cubic centimeters, which averages the capacity of five tablespoonfuls; then, for our percentages, six per cent. of one hundred is six, and one-tenth of one per cent. *is* one-tenth, and behold the thing is done! The example in full, then, is:

EXAMPLE III.—

For Miss I. J.

[*Apothecaries' System: six tablespoonfuls.*]

℞. Acidi Citrici.................... ʒ ij.
Olei Limonis .... ♏ ij.
Tere simul; dein adde
Aquæ....... .... f ℥ iv.
Solve, et adde gradatim
Potassii Carbonatis..q. s. ad saturandum.

[*Metric System: five tablespoonfuls.*]

	Gm.
℞. Acidi Citrici	6|00
Olei Limonis....................	0|10
Tere simul; dein adde	
Aquæ............................	100|00
Solve, et adde gradatim	
Potassii Carbonatis..q. s. ad saturandum.	

S.—" A tablespoonful as occasion arises."

K. L., M.D.,
No. 3, Third Street.

Office-Hours: 9 to 11 A.M.
August 27, 1894.

In the fourth example—the pills of blue mass, aloes, and rhubarb—the amounts are readily fixed. We want enough pills to last ten days or thereabouts; and the nightly allowance of each ingredient is about—apothecaries' system—two grains. A daily dose, then, of *two* pills, each containing *one* grain of the several constituents, will answer the purpose, and, plainly, twenty such half-strength pills will be needed. We must, then, order a mass composed of twenty grains each of the three constituents, and direct this to be divided into twenty pills. Metrically, we would rate the daily allowance of the several constituents at *ten* centigrammes, and the amount necessary for ten days would then be, of each, ten times ten centigrammes, or *one gramme*. The prescription would then read:

EXAMPLE IV.—

For Miss M. N.

[*Apothecaries' System: 20 pills @ gr. j. of each ingredient.*]

℞. Massæ Hydrargyri,
[Pulveris] Aloës Purificatæ,
[Pulveris] Rhei.............. āā ℈j.
Aquæ........................ q. s.

[*Metric System: 20 pills @ 0.05 Gm. of each ingredient.*]

℞. Massæ Hydrargyri,
[Pulveris] Aloës Purificatæ,
[Pulveris] Rhei..........āā 1.00 Gm.
Aquæ....................q. s.

M., et in pilulas no. xx. divide.
Or, M.: Fiat massa in pilulas no. xx. dividenda.)
S.—"Two nightly."

O. P., M.D.,
No. 4, Fourth Street.

Office-Hours: 8 to 9 A.M., 4 to 5 P.M.
August 28, 1894.

In example number five the amount is simply the number of pills, of a size and composition already determined by the Pharmacopœia, which the patient is likely to require for the present need. Assuming by the outlook of the case that the medication will probably have to be kept up for a week or two, and observing that six pills are used a day, the number *five dozen*, an allowance for ten days, suggests itself as a convenient round number to order. The prescription then becomes;

EXAMPLE V.—

For Miss Q. R.

℞. Pilulas Ferri Carbonatis.... ...no. lx.
S.—"Two pills, thrice daily."

S. T., M.D.,
No. 5, Fifth Street.

Office-Hours: 9 to 1.
August 29, 1894.

In the next example we want everything the same as in the last, except that the dose must be somewhat less—somewhere between two-thirds and three-fourths of the foregoing amount. We can accomplish the result easiest by giving fewer pills a day, but we can also, if we choose, preserve our method of giving two pills thrice daily by simply having each pill of but from two-thirds to three-fourths the pharmacopœial weight. For it so happens that this particular mixture the pharmacist, for pharmaceutical reasons, will probably compound afresh for each order, so that, having the pill-mass to make, it is just as easy for him to divide it into small pills as into large. We turn, then, to our Pharmacopœia, and find that each pharmacopœial pill weighs, in the metric system which alone is standard in the present revision of the Pharmacopœia, 0.30 gramme. For our sixty pills of standard weight the pharmacist, then, cuts up a pill-mass weighing 18.00 grammes. Let us order, now, a pill-mass of but *two-thirds* this weight, to be divided into the same sixty pills, and we shall have the diminished pill we seek. Such quantity will be, metrically, 12.00 grammes, each pill then weighing 0 20 instead of the original 0.30 gramme. For estimating by apothecaries' weight, we must calculate backward, so to speak, from the standard metric quantities of the Pharmacopœia. Twelve grammes is, roughly, three drachms, which quantity, divided into sixty pills, gives about three grains to each pill. We prescribe, then :

EXAMPLE VI.—

For Miss U. V.

[*Apothecaries' System : 60 pills @ gr. iij*]

℞. Pilularum Ferri Carbonatis...... ʒ iij.

In pilulas no. lx. dividendas.

[*Metric System : 60 pills @* 0.20 *Gm.*]

℞. Pilularum Ferri Carbonatis ... 12.00 Gm.

In pilulas no lx. dividenda.

S.—"Two pills, thrice daily."

W. X., M.D.,

No. 6, Sixth Street.

Office-Hours : 3 to 5 P.M.

August 29, 1894.

In the seventh example we are ordering a solution of silver nitrate of a certain strength for external use. Now, in medicines which, like silver nitrate, are employed in very different strengths of solution, we commonly rate the strengths, in apothecaries' system, by the number of grains to the fluidounce, speaking, by ellipsis, of a "five-grain solution," "ten-grain solution," etc.; while in metric measure we naturally speak of strengths in percentages, as a "one-per-cent. solution," "two-per-cent. solution," etc. In our present instance we will assume that we want *about* a fluidounce or so of a solution which shall be of *about* "twenty grains" strength. The amounts, then, appear in the stating. Thinking metrically, we should propose, probably, twenty-five cubic centimeters of a four-per-cent. solution, and the amount of the caustic to make such strength appears on the instant: for were the volume one hundred cubic centimeters, *four* grammes would, of course, give a four-per-cent. solution; but the volume wanted being but one-fourth this measure, one-fourth of four grammes becomes the weight of the silver salt to be taken. The prescription is, therefore:

EXAMPLE VII.—

For Self.

[*Apothecaries' System: "20-grain" solution.*]

℞. Argenti Nitratis.... ....gr. xx.
Aquæ Destillatæ..............f ℥ i.

[*Metric System: 4-per-cent. solution.*]

	Gm.	
℞. Argenti Nitratis	1	00
Aquæ Destillatæ....	25	00

Solve et S.—"For external use."

Y. Z., M.D.,
No. 7, Seventh Street.

Office-Hours: 2 to 4 P.M.
August 29, 1894.

These seven examples sufficiently illustrate the ways in which the data for computing amounts in prescribing commonly present themselves, and the methods of procedure in general in effecting the computation. There remains, however, a point wherein the beginner still stands in need of assistance. It is the problem presented where a basis is given in a *fluid mixture* under the common conditions that the total shall be an *exact bottleful*, as bottles are made, and that the dose shall be measured by the conventional spoonful. Under these conditions, knowing *about* how much basis we want for a dose, and *about* how many doses we are likely to require, what ready method is there for finding what round aggregate of basis to what even bottleful of mixture will give, to the right number of spoonfuls, the right amount of dose? To illustrate: We want to give, in fluid mixture, ten or a dozen four- or five-grain doses of some stuff: then on how big a bottleful shall we base our prescription; how big a spoonful shall hold our dose; and what bottleful and what spoonful will give a convenient round total of basis for a four- or five-grain charge per spoonful? To solve the problem, the first point to attack is the size of spoonful to take the dose of basis. If the basis is to be in *solution*, of course the factor of degree of solubility has a prime bearing; but assuming this not to stand in the way, then the next consideration is that, in *administration*, the strength of solution had better be limited to (round numbers, apothecaries' weight) *ten grains*, or (round numbers, metric weight) *fifty centigrammes* to the *teaspoonful*, and, to the *tablespoonful*, four times these weights, viz., *two scruples* and *two grammes* respectively. And in practice we even prefer—of course now speaking in a very general way—to give not more than *five grains* or *twenty-five centigrammes* in a *teaspoonful* of vehicle, nor more than *twenty grains*, or *one gramme*, in a *tablespoonful*. All these figures, however, refer to the concentration of dose as *actually administered*, and constitute limits of strength in the prescribing only when the dose is to be taken, without dilution, direct from the bottle. But it obviously may be a convenience, if the solubility of the basis permit, to order the mixture much more concentrated than the foregoing limits, with the understanding that it is to be properly diluted for the taking. Disposing, thus, of the first point, the second is, having the data of *dose of basis*, *number of doses*, and *size of spoonful*, to get that combination of totality of mixture and of basis which will practically fulfil the requirements of the data, on the one hand, and offer round numbers for convenience of prescribing, on the other. In the metric system, thanks to its simplicity, the matter is easily learned, for, since *any* aggregate of basis is equally easy of expression, we have but to bear in mind the number of spoonfuls to rounded metric volumes, as follows:

TABLE SHOWING NUMBER OF AVERAGE SPOONFULS TO ROUND METRIC VOLUMES.

Teaspoonfuls.	Tablespoonfuls.	Cubic centimeters.
5	..	25
10	..	50
20	5	100
40	10	200
50	..	250
60	15	300
80	20	400
100	25	500

In the table, only those equivalents are given which are likely to be calculated from in actual practice. To illustrate the application, let us assume that we want about a dozen or so doses of somewhere between eight and twelve centigrammes of a thing to be given in fluid mixture. From the smallness of the dose we naturally select a teaspoonful rather than a tablespoonful for the measure of vehicle to hold each of the same. Then referring to the table, and seeing that a fifty cubic centimeter measure offers an aggregate of ten teaspoonfuls, we at once select such total for our mixture, and calculate the total of basis on the scheme of ten doses. And in such calculation appears the enormous advantage of the metric over the apothecaries' system. For now *any* total of basis is equally easy of expression and of actual weighing out, so that we are enabled, untrammelled by any inconveniences in calculation or expression, to prescribe our doses *exactly* as per the apparent therapeutic indication. Thus, in the present instance, we are ordering an aggregate of ten doses ; now let the estimated dose be any of these several weights —*eight*, or *nine*, or *ten*, or *eleven*, or *twelve centigrammes*—and the aggregate is equally easy of calculation and expression : we have but to multiply by ten, and set down in ordinary decimal notation, respectively, thus: 0.80, 0.90, 1.00, 1.10, 1.20 Gm.

If, however, we use the apothecaries' system, we are practically bound by its clumsiness, in the way that odd amounts are so inconvenient of expression as to be impracticable for prescription. We have in this case, then, to learn, by rote, certain set combinations of aggregates for mixture and basis, and to figure our dosage no closer than can be done by those combinations. Such combinations, for single doses of from five to twenty grains, are exhibited in the following table. For minute doses, the total being easily expressed in grains, the difficulty now referred to does not obtain.

In the table on page 178, as in the foregoing, combinations that yield awkward amounts are omitted mention, and the calculation is on the basis that obtains in the case of modern spoons, of *six* teaspoonfuls and *one-and-a-half* tablespoonfuls to the fluidounce. But if it be known that an old-fashioned small-sized spoon is to be used, or if the patient use a graduated measuring-glass, then the calculation should be on the scheme of *eight* teaspoonfuls and *two* tablespoonfuls, respectively, to the fluidounce—should, that is, rate the teaspoonful as a fluidrachm and the tablespoonful as half a fluidounce. In such case the table on page 179 applies instead of the preceding.

These tables are offered simply for reference until the beginner, by practice, comes to remember, as he soon does, the most handy of the combinations.

So, having studied the how to compose, how to write, and how to compute amounts for a prescription, it remains but to note a something still to be done, even after the prescription is ready for delivery. And that is, before such delivery, critically to *review* the paper—scanning deliberately drug-names, amounts, and doses. Found early, an error is a matter of the stroke of a pen; found late, perhaps of a coffin and a coroner's jury!

Table showing Number of Average Spoonfuls to Round Apothecaries' Volumes, and Amount of Basis to yield Doses to the Spoonful, of 5, 10, 15, 20 Grains severally.

Teaspoonfuls.	Fluidounces.	Total of basis, in order to give to the teaspoonful, severally—			
		Five grains.	Ten grains.	Fifteen grains.	Twenty grains.
3	½	gr. xv.	ʒ ss.	....	ʒ j.
6	1	ʒ ss.	ʒ j.	ʒ iss.	ʒ ij.
12	2	ʒ j.	ʒ ij.	ʒ iij.	℥ ss.
24	4	ʒ ij.	℥ ss.	ʒ vj.	℥ j.
36	6	ʒ iij.	ʒ vj.	ʒ ix.	℥ iss.
48	8	℥ ss.	℥ j.	℥ iss.	℥ ij.
72	12	ʒ vj.	℥ iss.	...	℥ iij.

Tablespoonfuls.	Fluidounces.	Total of basis, in order to give to the tablespoonful, severally—			
		Five grains.	Ten grains.	Fifteen grains.	Twenty grains.
3	2	gr. xv.	ʒ ss.	....	ʒ j.
6	4	ʒ ss.	ʒ j.	ʒ iss.	ʒ ij.
9	6	..	ʒ iss.	....	ʒ iij.
12	8	ʒ j.	ʒ ij.	ʒ iij.	℥ ss.
18	12	ʒ iss.	ʒ iij.	...	ʒ vj.

TABLE SHOWING THE NUMBER OF FLUIDRACHMS AND HALF FLUIDOUNCES TO ROUND APOTHECARIES' VOLUMES, AND THE AMOUNT OF BASIS TO YIELD TO THE FLUIDRACHM AND HALF FLUIDOUNCE DOSES, SEVERALLY, OF 5, 10, 15, 20 GRAINS.

Fluidrachms.	Fluidounces.	Total of basis, in order to give to the fluidrachm, severally—			
		Five grains.	Ten grains.	Fifteen grains.	Twenty grains.
4	½	℈j.	℈ij.	ʒj.	℈iv.
8	1	℈ij.	℈iv.	ʒij.	℈viij.
16	2	℈iv.	℈viij.	℥ss.	℈xvj.
32	4	℈viij.	℈xvj.	℥j.	....
48	6	℥ss.	℥j.	℥iss.	℥ij.
64	8	℈xvj.	. ..	℥ij.	....
96	12	℥j.	℥ij.	℥iij.	℥iv.

Half Fluidounces.	Fluidounces.	Total of basis, in order to give to the half fluidounce, severally—			
		Five grains.	Ten grains.	Fifteen grains	Twenty grains.
2	1	gr. x	℈j.	ʒss.	℈ij.
4	2	℈j.	℈ij.	ʒj.	℈iv.
8	4	℈ij.	℈iv.	ʒij.	℈viij.
12	6	ʒj.	ʒij.	ʒiij.	℥ss.
16	8	℈iv.	℈viij.	℥ss.	℈xvj.
24	12	ʒij.	℥ss.	ʒvj.	℥j.

PART II.

TECHNOLOGY OF MEDICATING.

CHAPTER I.

MODES OF MEDICATING.

THE possible effects of medication are of two kinds: first, the effects upon the tissue with which the medicine comes in contact, produced directly by virtue of such contact; and, secondly, indirect *consequences* of such effects, appearing, it may be, even in distant parts, either as nervous reflex phenomena, or as consequences of changed blood-supply, etc. Therapeutically, either or both of these sets of effects may be of importance, but in our present technological study we evidently have to do only with the first consideration, namely, the matter of the *bringing of medicines into contact with tissue*, which procedure is the prerequisite for all medicinal effect, immediate or remote.

The subject naturally divides itself into two parts, viz., how to touch with a medicine the *surfaces* of the body, on the one hand, and how, on the other hand, the *underlying tissues*.

Of the surfaces, the *skin* is so obviously accessible to any mode of medication that, in its case, but few technical points present for discussion. Of these, the first to note is that the skin, because of its comparative insensitiveness and resistance to transfusion, can safely bear, even of poisonous things, far stronger and more extensive applications than can the mucous membranes. Yet, as will be seen in detail further on, the skin *can* absorb, so as to charge the blood with the absorbed thing to a dangerous or even a fatal degree. Powerful medicines, and especially those that combine the qualities of *potency*, *volatility*, and *high diffusion-power*, as, for instance, *carbolic acid*, must therefore not be applied to the skin too strong or too extensively. Secondly, it must be remembered that the skin is a true organ, having physiological functions; and that applications of a kind and extent to interfere seriously with the performance of such functions are, for that reason, inadmissible. The persistent covering of nearly the whole of the skin with an impervious layer of ointment—even if the ointment be in itself innocent—is therefore an unadvisable proceeding.

Thirdly, it is to be noted that, in medicating the skin, the medication will be the more intense the *cleaner* the surface. Soap and water, or more strongly alkaline lotions, to remove dried grease and epithelium, are, thus, important applications to precede a course of skin medication. Fourthly, when hairy parts are to be medicated, the hair should first be cut short or the part shaved. And, fifthly, it is to be remembered that, of the vehicles into which medicines are put for application to the skin, fatty matters penetrate cracks and crannies better than aqueous fluids, and, of the fatty matters, *oleic acid* and *oleates* dissolved in excess of the acid are by far the most penetrating.

The other exposed surfaces of the body besides the skin are, of course, the *mucous membranes*. Concerning these in our present consideration, we note at the outset two important points. The first is the matter of the very different *degree of accessibility* of the different mucous surfaces. Some, such as those of the eye and mouth, are practically as freely exposed as the skin itself, while others, in varying degree, require special methods, and even special appliances, in order to be reached in medication. And the second point, and an important one, is that of the equally different *degree of sensibility* of the different mucous membranes. Some, by the conditions of their functions, must normally suffer touch from without, while to others foreign touch is obnoxious. The former must necessarily be insensitive, while the latter, finding a safeguard in sensitiveness, may be exquisitely tender. And, of course, the present interest in these facts is the plain indication that, in medicating, applications to mucous membranes must be proportioned in strength to the natural sensibility of the part touched. The most sensitive mucous surfaces are those, severally, of the *cornea*, the *upper* portion of the *nasal cavity*, and the *larynx;* next come, in order, the general surface of the *conjunctiva*, the *air-passages* beyond the larynx, the *middle ear*, the lower portion of the *nasal cavity*, and the *urethra;* while least sensitive are the mucous coverings of the *alimentary canal* and of the *female generative organs.*

Of the means of reaching these various parts, we note that the *conjunctiva* is immediately accessible; the only technical point to make being that, to secure thorough application to the retro-tarsal fold, the upper lid must be fully everted, while the patient is directed to look strongly downward. Otherwise the very part that in conjunctival diseases most needs medicinal touch will escape the application altogether.

The mucous membrane of the *nasal cavity* is very difficult of thorough access, and the cavities that communicate with the nose by small apertures are practically wholly beyond reach. The nasal cavity may be medicated by the snuffing-up of dry powders or fluids—an imperfect method—or powders may be blown up the nostrils by a rubber bag with a nozzle, or fluids may be injected either from before or behind. A danger of such injections, now happily pretty thoroughly appreciated by practitioners, is that the fluid of the injection may find its way *via* the Eustachian tube into the *middle-ear*, a cavity whose mucous membrane will almost certainly resent such intrusion by inflaming. The *nasal douche* of a few years ago is now, therefore, very generally condemned, and even the *posterior nasal syringe* falls under the ban with many. This latter appliance is certainly the least likely to offend, if nasal injection is to be practised at all. The posterior nasal syringe is simply a syringe with a nozzle long enough to reach through the mouth to the fauces, and with the end of such nozzle upturned and so pierced with holes as to throw the fluid *backward* in relation to itself, but that is *forward* in relation to the nasal cavity, *i.e.*, from the rear opening thereof forward toward the nostrils. The curve upward at the end of the nozzle should be quite sharp—sharper than the instrument-makers generally give ; the syringe being made of hard rubber, the nozzle may gently be warmed, and then the proper curve easily given to it. For *self*-use it is also a convenience to bend body and nozzle, where they join, to a right angle. The body then hangs vertical during application, an easier position for the self-injector.

The *Eustachian tube* is reached by the Eustachian catheter, an instrument whose application belongs to the domain of surgery.

The *mouth* is, of course, directly accessible, and the *palate* and *pharyngeal cavity* practically so. To the posterior portions of the oral cavity the method of *gargling* applies, but the same is a very ineffectual procedure, the pharyngeal cavity proper being scarcely touched at all by the fluid. Applications of *spray* are here peculiarly happy.

The *larynx* can be reached by proper probang, guided by a view in the laryngoscope-mirror, but such special and delicate manipulation, of course, must be taught to the individual clinically. Gaseous medicines or fluids in fine spray can be applied by inhalation—remembering that, of course, such only are allowable for this application as are *innocent* and *unirritating*.

The *air-passages beyond the larynx* are, obviously, locally accessible only through inhalation. Gaseous medicines of the kinds just indicated can thus be applied, but there are few of such kinds that are of much value as local respiratory remedies. *Fluids in fine spray* can be inhaled, of course, if not too irritant, but there is much reason to doubt if the spray penetrates very far beyond the larger bronchi.

The *bladder* can be reached by injection through the urethra; and the *urethra* itself by injection, by the insertion of a medicated plug of cacao-butter (urethral suppository "bougie"), or by the passing of a sound smeared with the medicament. The instruments and manipulation here required are, again, as in the case of the larynx, too special to attempt to treat of didactically. The urinary passages can also, in the case of certain volatile oils and resins, such as copaiba and cubeb, be reached by impregnation of the *urine* with the medicine through the roundabout way of *swallowing* the drug. In such case the drug is absorbed into the blood, and, being excreted in the urine, comes thus locally to be applied to the urinary mucous membranes.

The *vagina* can easily be syringed, and the patient herself taught the procedure. Nothing more special is required than a long nozzle, perforated at the extremity with holes delivering in all directions, which nozzle can be applied to any form of syringe. The *Davidson* type is convenient, but handier yet is a rubber bag with long, flexible rubber tube ending in the nozzle, and armed with a stopcock of simple device. The bag is filled with the injection, then hung against the wall on a nail three or four feet above the level of a seat. The nozzle is inserted (the patient, of course, sitting over an ample receiving-vessel), the stopcock is turned, and gravity then determines a steady flow of the injection, the *force* thereof being the greater the higher the bag is hung. During a vaginal injection, the nozzle should be rotated from side to side, and withdrawn and pushed up from time to time, so as to secure irrigation of all parts. More certain for this end, although more inconvenient, is the plan of having the patient on her knees and elbows during the injection. In that position the walls of the vagina tend to fall asunder, and the injection thus more surely reaches every point of surface. The vagina can also be reached by vaginal suppositories, or by medicated pessaries, or by instrumental appliances under exposure by a speculum.

The *uterus* can be injected, but at great risk of having the injection escape into the peritoneal cavity through the Fallopian tubes, with possible mischievous or disastrous consequences. Otherwise, medicinal applications to this cavity are made by special instrumental means.

The *stomach* and *intestines* are reached by swallowing, or, so far as concerns the stomach, by the *stomach-pump*, an apparatus with a flexible tube to reach down the œsophagus into the cavity of the stomach. By an arrangement of valves, fluids can be pumped *into* the organ, as well as the organ's previous contents pumped *out*. In medicating the alimentary tract, it must be remembered that here is one of nature's greedily *absorbent* surfaces, and that the majority of things put into the canal for local effect cannot be prevented from also finding their way into the blood. Some medicines, however, either because of insolubility or of low-diffusion power, are so slow and imperfect of absorption that quite a valuable local effect can be produced by them in the bowels without the system at large being affected. Notable examples in point are the salts of bismuth and saline purges. We must also see to it that in our eagerness to medicate this tract we do not interfere with its functions, as by spoiling appetite, exciting nausea or diarrhœa, etc.

The *rectum* may be medicated by suppositories or by injection. In giving a medicinal enema, the points should be observed first to inject plain water in sufficient volume—a pint or more—thoroughly to wash out the cavity from fæcal matter. This injection having been discharged, with its washings, the bowel is given a short rest, and then the medicated enema is to be *slowly* injected. And such enema should be of small bulk—not over two fluidounces—and blood-warm, so as not to excite the bowel to its expulsion. On withdrawing the nozzle, the fingers or a napkin should be pressed against the anus for a few seconds, and the patient, if old enough to understand, cautioned to resist any inclination to strain. In practising any rectal injection, the points should be observed to have the nozzle *warm* and *well greased*, and to pass it, after it is once engaged within the sphincter, *upward* and *backward*, following the concavity of the sacrum. The passage should be slow, and, in the case of a crying child, the pressure should be exerted only during the *inspirations*, when the abdominal walls are relaxed.

In thus running over the special means of medicating the mucous membranes, a *fine spray* of fluid has several times been mentioned. This condition of a fluid medicine is a valuably convenient one, since it allows fluids to pass without irritation into the air-passages, and, even to exposed parts, enables us to apply liquids thoroughly and evenly without *drip* and *slop*. Fine spray is obtained by appliances based on the following simple principle: let a rather narrow tube, with one end drawn to a fine orifice, have the other end, open, immersed in a vessel of fluid. Then close to, and at right angles to, the free fine orifice, let there be a similar fine orifice of a *second* tube, through which from behind a strong blast of air or steam can be driven. By the partial vacuum caused by such blast, the fluid from the vessel is sucked up the first tube and appears at the fine orifice thereof. But no sooner does a drop thus show itself than it is, at once, literally *blown to atoms* by this same blast through the second tube —is, in short, dispersed in a cloud of excessively fine spray. The contrivances operating on this principle are called *spray-producers* or *atomizers*, and are modelled of different shapes to suit different special applications. The blast is commonly obtained by hand-pressure on a valved rubber sphere connected with the apparatus by a flexible tube bearing a second rubber sphere midway in its course. This mid-sphere acts as a reservoir, determining a steady blast during the intermittent action of the terminal sphere. Or, should an intermittent blast only be wanted, the mid-sphere may be squeezed and the blast will now only take place during the continuance of the pressure. Where a prolonged application is needed, as in spray-inhalations, the blast is most conveniently obtained by steam from a small boiler, special apparatuses for such end having been devised under the name of *steam-atomizers.*

To apply medicines to parts *beneath* the surface, meaning all parts of the body save skin and mucous membranes, we can for a few special purposes inject into muscles or into serous cavities, but in the enormous majority of instances we medicate all underlying tissues by putting the medicine, by some means, into the *blood*, thus, as it were, shipping it to its distant destination through that universal avenue of communication. But in so doing arises at once a consideration which does not obtain in surface-medication. It is that we cannot here, as we can there, restrict the contact of the medicine to the part required to be medicated; for the stuff, being dissolved in the general mass of the blood, must, perforce, go wherever the blood goes; we cannot confine its tour in the vascular system to any one artery going to a particular part. To medicate brain, or spinal cord, or kidneys, or a single neuralgic nerve even, there is no help for it, but we must bathe the whole blood-supplied organism with medicated blood, and thus, perhaps, secure our therapeutic result over one part at the expense of considerable annoying derangement of others. This is unfortunate, of course, but it is unavoidable, and all we can do is to give preference, among drugs of similar therapeutics, to the one that happens to work the maximum of a therapeutic effect with a minimum of extra-therapeutic disturbance.

Now, to *get* a medicine into the blood there are a variety of ways. The easiest, most natural, and therefore commonest method is to follow nature's course in getting *nourishment* into the blood, namely, to let the medicine be swallowed, and so be absorbed through the veins and lacteals into the general circulation. Not only is this the easiest way, but it has also the advantage over all others that substances in all conditions of crudity and of obstinate insolubilities can thus be dumped, as it were, upon the organism, and it will be the exception if the complex chemistry of the *primæ viæ* will not extract the virtues of the dose in soluble form and duly deliver them over into the blood-vessels. Yet still, for many reasons, the stomach may be objectionable or even absolutely unavailable as the avenue through which to enter the vascular system. This same deranging tendency of drugs just spoken of may, and frequently does, show itself locally upon the stomach, and loss of appetite, or nausea, or even vomiting, may be the cost of forcing a drug into the system by this means. Or even if the medicine be, by rights, innocent of such tendency, the stomach may be in such morbid state from disease that even normally wholly harmless things, like ordinary foods, upset it and are not to be borne. Or, because of corrosive poisoning or of stricture of the œsophagus, the organ may be absolutely disqualified for use ; or it may practically be so because its absorbing capacity is in abeyance from inflammation of its mucous membrane, from narcotic poisoning, or from general collapse of all vital powers when life is at low ebb from serious disease. Thus when an individual is in profound coma from opium-poisoning or is in the collapse of cholera, absorption by the stomach stops, and it is worse than useless to thrust drugs into the paralyzed organ with a view to their absorption. Or, though the stomach may bear a drug fairly well, and its functional activity be in good state, yet it must be remembered that the rate and thoroughness of absorption by this organ are necessarily subject to variation. The potent gastric juices may chemically attack the medicine and thus defeat our purpose, or the mere mechanical obstacle arising from the presence of a large mass of food, with which the drug necessarily becomes mixed upon swallowing, may so delay full absorption as to be of serious cost. Hence, when a therapeutic call is urgent and an effect of dosage both *prompt* and *full* is imperative, we dare not risk the uncertainties of the stomach, but prefer other methods.

Now, when, for any of the above reasons, we seek other methods, the most natural avenue after the stomach is the *rectum*, and general medication by enema or suppository is not uncommon. But apart from the obvious inconvenience, and even, it may be, indelicacy, of using this approach, the method has its disadvantages. The absorbing power of the rectum is not so great as that of the stomach, and, more particularly, we have not here the complex digestive fluids that so readily, in the stomach or small intestines, attack crude or insoluble drugs and reduce them to a form capable of absorption. Hence to introduce a medicament into the blood by means of the rectum we commonly, in the first place, administer twice as much as we would give by the stomach, and we see to it, in the second, that the substance is, primarily, either dissolved or is in condition to undergo easily simple aqueous solution.

Besides the rectum, the lungs afford a natural avenue of approach to the vascular system, and offer, for the present purpose, the advantages that they *absorb* quickly and thoroughly. But, obviously, the medicines that can be given by inhalation are limited to such as are highly volatile and at the same time respirable. *Nitrogen monoxide* gas, *ammonia*, and certain volatile ethereal fluids, such as *amyl nitrite*, *ether*, and *chloroform*, are pretty much the only things given by this method.

Next, we can avail ourselves of the *skin* for purposes of constitutional medication, and that, too, in a variety of ways. Certain easily diffusible substances, in solution, will be absorbed through the sound skin if only laid thereon, by wetted cloths; but such means is so crude, and dosage so uncertain, that the procedure is nowadays rarely resorted to.

A method of determining absorption through the skin that is, however, much used, and with great advantage with the particular drug *mercury*, is to rub into the skin a fatty preparation carrying the medicament, when, largely by mechanical forcing, the particles find their way into the texture of the skin, and thence, after undergoing chemical conversion into a soluble compound, into the blood. Other medicines, also, can be introduced in this way, such as, for instance, the alkaloids ; but since in the majority of cases this class of substances can be given better by hypodermatic injection, this method of *inunction*, as it is called, is, in the case of alkaloids, rarely employed. Then—again practically confined to preparations of mercury—the drug can be *sub-*

limed and the vapor allowed to condense upon the skin, when absorption will ensue.

Next, passing from natural ways, we can put medicines into the blood by *artificial methods* through surgical procedures. We can, in the first place, by appropriate apparatus, pierce a vein with the nozzle of a syringe, and thus inject directly into the vascular canals. Such method is occasionally used by practitioners of great hardihood, even in the case of powerful drugs, but is to be condemned for its obvious dangers, and, by the bulk of the profession, is resorted to only as a means of introducing into the blood bland nutriments, such as milk or blood, or a simple saline solution—all as a means of keeping the heart going when a patient is in desperate straits. Another method, less severe, is to raise a small blister upon the skin somewhere, remove the separated epidermis, and, upon the raw surface beneath, lay the medicine; which, in this case, must, for obvious reasons, be one which is, at the same time, efficient in small dose, readily soluble, and not unduly irritant. This method, the *endermatic*, has, however, been entirely superseded by the far better, though physiologically similar, method by subcutaneous or *hypodermatic* (barbarously miscalled *hypodermic*) injection. The procedure in this case is simply to pierce the skin with a fine and sharp nozzle of a small syringe, and then inject into the loose subcutaneous connective tissue the medicine—of course in solution. From the purely physiological point of view this is the best method of all. Absorption is rapid, thorough, and almost invariably certain, under all conditions and circumstances of the patient's morbid state; derangement of digestion is reduced to a minimum, and, with some drugs, certainly, the therapeutic effect is more intense, or more persistent, or even, as in the case of morphine given to combat neuralgic pains, more *radical*, than where the same drug is administered by other methods. Because of these great advantages the hypodermatic syringe is as universal a tool with the practitioners of to-day as was the lancet with our fathers, but yet the method has its restrictions. Not everything can be given by subcutaneous injection. Obviously the medicament must not be severely irritant, else great pain, and even subsequent inflammation and abscess at the seat of puncture, will result. Again, if a solid, the drug must be soluble in reasonably bland fluids, such as water, weak alcohol and water, glycerin and water, or solutions of mild salts; and, lastly, it must be a thing whose *dose* is small enough for the bulk of the injected fluid not to exceed two cubic centimeters (thirty minims)—indeed, much less than this measure is to be desired. Then, again, the procedure itself is often seriously objectionable. Simple though it be, and insignificant though the pain, yet to sick, nervous, and excitable women and children the idea of being stabbed, though ever so delicately, is terrifying. Then, except where well-trained nurses are in attendance, the physician must himself administer the dose—often an obvious great inconvenience to physician and patient both, and to the patient also, it may be, an ill-to-be-afforded expense. With the one medicine, moreover, most commonly given by the hypodermatic method, namely, some salt of morphine, there is often the serious danger that the "opium habit" may thus unwittingly be begun, and, if so, it will be the particular form of this enslaving vice that will work most damage and be the most cruelly hard from which to break away. The dainty poniard of the "hypodermatic" is, then, not always the boon that many young practitioners of to-day—to judge from their freedom therewith—would seem to consider it.

To *give* a hypodermatic injection, first have a *good syringe.* The piston should work easily and evenly and without leakage, which it probably will *not* do if the cylinder be of *glass*, but *will* if it be of silver, celluloid, or—if the workmanship be good—of hard rubber. There must, next, be a *graduation* somewhere to tell the amount injected. This, in the case of glass cylinders, may be on the cylinder itself, but, in the case of instruments made of opaque material, must necessarily be on the piston. In the latter case there is commonly a small screw-collar on the piston, which, by setting at the proper mark, stops the piston from going beyond a certain distance, and so limits the amount of injection possible to deliver. On purchasing a syringe, the graduation should be tested for accuracy before trusting to its possible false showings. The *needle* should be clean, sharp, and free from rust—conditions best maintained by having the needle of gold, with, of course, a point of harder metal. The point should be a plane bevel, and the whole needle should be *fine* rather than coarse.

Having, thus, a good syringe, have, next, a *good solution*. No *dirt*, no *decomposition*, and no *free acid* must find place therein, and we must be *certain of the strength*, which, by the way, must not be too great. Water is the best vehicle for the injection, and things soluble in that fluid are therefore the favorites for administration by the hypodermatic method. The solution is better made fresh, and any clean water fresh from a tap is better than stale and, therefore, almost certainly *mouldy* distilled water. If solutions be kept, assuming them to be salts of alkaloids, they must be charged with some preservative, such as hydrate of chloral, carbolic acid, salicylic acid, etc. One per cent. addition of any of these bodies proves antiseptic, but the same are all more or less irritant, and hence fresh solutions are preferable. For convenience in making fresh solutions, manufacturers offer tablets charged with fixed quantities of the things commonly used for hypodermatic injection. These tablets are simply dissolved in a few drops of water on the occasion of the injecting. Such tablets are made of *gelatin* or of *sodium sulphate*, the salt in the latter instance being given form and cohesion by powerful pressure. These medicated tablets, if of reliable make, are exceedingly convenient, the compressed tablets of the sodic salt especially so, provided they are fresh enough to dissolve readily.

Being ready with a good syringe and a good solution, we fill the one with a sufficiency of the other, then hold the syringe vertical, needle-end up, and gently push upon the piston until fluid appears at the needle orifice. Thus the bubble of air, which it is practically impossible to prevent from having place within the cylinder, is discharged, and we are now certain that the syringe is just as full of solution as it purports to be. We next *fix the dose* beyond possibility of misadventure by observing where the piston stands in relation to the graduation upon its shank, and by then running down the screw-collar so far that on driving the piston home as far as the collar will allow it to go we shall inject just the desired quantity. Then a fold of skin is pinched up with the fingers of the left hand, and into the triangular slope trending downward from between the fingers the needle is quickly plunged, in direction carefully parallel to the surface of the limb beneath. After having been pushed in to a depth of from half to three-fourths of an inch, it should be gently withdrawn a trifle, and worked slightly from side to side. Next the injection is made, rather slowly, after which the nozzle is quickly pulled out, and, as a matter of precaution against leakage, a finger is lightly pressed for a few seconds on the skin puncture. The only danger in the procedure is the possible pricking of a vein, thereby throwing the injection directly into the blood-current. But by observing the rules just given, the chances of this accident are very small indeed.

As regards the site of the injection, where, as is most commonly the case, the aim is simply to get the medicament into the general blood-supply, the site is indifferent and may be selected according to convenience. Situations are preferred where the skin is lax, thin, and free from hair; and since these conditions are fulfilled on the arm —also a part handy of access—this is the most frequently selected site. Where, however, the injection is meant to be, in part at least, of direct local effect, then, of course, the little operation is practised as near the affected spot as may be. Yet, under any circumstances, a thick, closely adherent portion of skin, like the scalp, should be avoided. The *dose* by the hypodermatic method should invariably be less, even by one-half, than that which would be given by the stomach to produce an equal effect.

CHAPTER II.

DOSAGE.

OUR last topic in general technology is *dosage*, meaning matters connected with the *determination* of doses of medicines. Here we have, first, to consider certain principles that apply generally, and, secondly, to note the circumstances under which ordinary doses must specially be modified. The general principles of dosage are best studied by examples. Let us first suppose a simple case: A woman is faint, and we medicate to whip up the faltering heart. A teaspoonful of brandy circulating in the blood will ordinarily do the business, and a teaspoonful of brandy is thereupon prescribed and taken, with, we will suppose, the expected effect. Here the matter has been simple; the need for a medicinal influence was transient, and, there being no objection, the full quantity of drug required for the effect was given at once—here, that is, *dose* has been made to equal quantity necessary to be present within the system at a given time. But let us take another case: As a teaspoonful of brandy in the blood will oppose emotional heart-failure, so will twenty grains of quinine oppose an expected paroxysm of ague.

Shall we, then, in such circumstance, give the quinine as we gave the brandy—all at once? Better not, if we can avoid it, for while in the stomach *in transitu* to the blood, such a considerable dose will be likely to nauseate. And we *can* avoid it, for the reason that this medicine, once absorbed, stays in the blood several hours, so that, the time of the disease-onset being known beforehand, we have hours at our disposal for the medicating. So we do what the military commandant does who must garrison with a strong force a fort whose approach is a weak bridge—*we take time, and march the command across in small detachments;* we break up, that is, our heavy charge of quinine, and give it, as the phrase is, in *divided doses*—four grains, say, every half-hour until five doses shall have been taken. The result is practically the same as if the whole had been given at once—the full garrison is present when the enemy attacks, and the assault is foiled. Next, a third case, where the conditions differ—conditions that may be paralleled in our illustration of the fort in this wise: The fort is to be garrisoned, not to resist a passing single assault merely, as hitherto instanced, but a *prolonged siege.* At once a new factor enters into relation. *Desertions* are frequent and inevitable, which, unless offset by new enlistments, will in time dissipate even the entire command. So in medicating, the avail of the remedy, perhaps oftener than not, must persist in full force through quite a siege, while yet the molecules of the dose on duty are steadily deserting by excretion. *Reinforcement*, then, plainly presents now as a feature of dosage, and the practical question at once arises by which of two opposite methods shall such a reinforcement be carried out. Shall we, as it were, by frequent single recruiting, fill vacancies as fast as made, or shall we deliberately wait until the depletion be considerable, and then, at a stroke, restore to full quota? Evidently the latter method is the more convenient, but evidently yet it fails of the effect which the other secures of *maintaining the strength of the command steadily at a fixed figure.* And herein is the pith of the whole matter—a consideration which only rather recently has received the thought it deserves, namely, the importance of having the remedial impression, while it lasts, *equable.* The advantages of such equability are obvious, and often indeed the medication may wholly fail of its end unless this condition be fulfilled. Now, such fulfilment is only possible by *frequency of reinforcement*, and thus obtains the important rule that, in continuous medication, after the system is once properly charged with the drug, *renewals are to be on the plan of "little and often," rather than on that of large doses at long intervals.* But in obeying the rule two points are to be considered: First, that the actual frequency of renewal will *vary greatly in the case of different drugs*, according to the persistence of their effects on the one hand and the rapidity of their elimination on the other. Thus the heavy metals are, so to speak, tenacious upon the tissues; their effects are prolonged, and their excretion slow. Such an influence, therefore, as constitutional mercurialization can be maintained at an even pitch by renewals not oftener than thrice, twice, or even, by certain methods, once daily. On the contrary, flitting principles that swoop on the wing, as it were, such as ethers and many alkaloids, must be repeated, in dosage, with great frequency. If a heart is to have its pulse-rate evenly depressed by aconite, the renewals must be at least hourly, otherwise the pulse-rate will rise and fall in a regular wave between dosings. The second point to be considered is that, in cases where great frequency of renewal is theoretically indicated, we must yet hold our hands somewhat, lest, though we secure the desired evenness of remedy-action, we kill our patient by the very so doing! For in serious illness every disturbance of the sufferer tends to exhaust, and even so seemingly small a matter as the giving of a powder or a potion taxes appreciably the strength. In such condition, therefore, the wise physician recognizes that *extreme* scientific precision of medication is not worth the worry it entails.

We have thus elucidated the following as general principles of dosage in constitutional medication: The *basis of calculation is the percentage of drug to blood necessary for the effect*, a quantity which, under ordinary circumstances (exceptions anon), is for each drug a fairly fixed quantity. Then the dose, proper, follows thus: If the effect need be but transient, and if there be no objection, the whole amount necessary to establish the percentage is given at once, as in the case of the teaspoonful of brandy to revive from a faint; if, however, the need still being but a passing one, there be an objection to a large single dose, as in the quinine example, the quantum is given in fractional parts at proper intervals. If the influence must be at all prolonged, the requisite percentage is first established by either of the foregoing methods, and then maintained by reinforcements made, preferably, small and frequent, the dose at each renewal being, of course, duly proportioned to the frequency.

The next point in dosage is a simple and obvious one. One grain of quinine appetizes; twenty grains derange digestion but develop the new potence of reducing fever-heat. Both effects are utilizable therapeutically, but the *dose* for the different purposes differs enormously. With drugs of manifold therapeutic powers, therefore, *purpose* is an essential factor in estimating dose, the same drug having literally two or more "doses" according to the effect sought.

Such are the principles that regulate dosage in general, and we pass now to the independent consideration that even with the same drug, given in the same way and for the same purpose, the dose is not always the same. This may be because of a variety of reasons. In the first place, plainly, in constitutional medicating, bulk of remedy must bear relation to bulk of patient. The basis being, as we have seen, the establishment of a given percentage of drug to blood, to affect equally a two-hundred-pound oarsman on the one hand, and his weazen ninety-five-pound coxswain on the other, will take quite different dosage. And in prescribing for *children* this consideration of size of patient is plainly of particular moment, doses needing to be scaled systematically to the dimensions of the recipient. A practical point, then, is to obtain some formulæ by which this same scaling can readily be done. We have such, but before presenting them it is wise to remind that with children other conditions than mere size may influence the effect of a dose, so that in the cases of certain drugs, and even of certain classes of drugs, the scaling according to body-weight will require special further modification. Of general formulæ, however, so far as they apply, we have several, of which two, because of their ease of application, are the favorites, wherein, for convenience' sake, the dose is related to the age, simply. Of course all children of the same age are not of the same bulk, but yet for purposes of a general formula it is accurate enough to assume them so, any considerable departure from average size in a particular case being easily allowed for after the average dose has been obtained. The formulæ are as follows: *Young's* rule is, that, taking the adult dose at unity, the fraction thereof proper for a child of given age may be found by the formula:

$$\frac{\text{age}}{\text{age} + 12}$$

At age six, for instance, the fraction is $\frac{6}{6+12} = \frac{6}{18} = \frac{1}{3}$: *i.e.*, a child six years old takes one-third of the adult dose. *Cowling's* formula is, under the same premises:

$$\frac{\text{age at next birthday}}{24}$$

At present age six, that is, the formula gives the fraction $\frac{7}{24} = \frac{1}{3} -$: rather less than one-third. In general, with the younger ages, Cowling's formula yields a slightly smaller dosage than Young's; the one or the other may be used, therefore, according as we may wish scant or full dosing for the age.

But, apart from matter of size, there are many conditions—conditions of patient or of his environment—which may distinctly modify drug-influence, enhancing, enfeebling, or distorting, as the case may be. When such conditions obtain, doses must evidently be changed to suit. The more prominent of the conditions and their effects are as follows: First, *age*. *Children* are, as a rule, more susceptible to drug-influences than adults, though with a few drugs the reverse obtains; and particularly with children, and with old people also, actively perturbing or depressing effects are badly borne. *Sex* gives similar results, women being more impressible than men. *Climate* has an influence, again, warm weather determining, in general, disproportionate exhaustion after violent or debilitating therapeusis, and, in particular, undue susceptibility of the *digestive* apparatus to disturbing measures. *Custom*, *i.e.*, continued taking, with some drugs enhances, with some enfeebles the effects, with some modifies the intensity of certain of the effects only, and with some has no influence. One of the most prominent instances of a modifying influence from custom is in the case of narcotics, typified by opium and alcohol, where the more obvious functional derangements of the nervous system become proportionately less and less in habitual indulgence, while, at the same time, the subtler nutritive changes induced—degenerations and low inflammations—march on in full measure. *Individual idiosyncrasy*, like custom, may work either way, curious instances of exceptional susceptibility on the one hand, and of insusceptibility on the other, presenting themselves from time to time. With certain drugs individual idiosyncrasy as regards their influence is peculiarly common, necessitating exceptional caution in their prescription to stranger patients. Tobacco offers a well-known instance of this peculiarity, while, of drugs proper, opium, ipecac, and mercury afford marked examples.

Next, *special physiological status* of the system, generally, or of some part concerned, often affects, and most profoundly, the influence of a drug. Thus, locally, a dirty, thick, or inflamed skin will absorb less perfectly than a clean, a thin, or a healthy one ; a full stomach will be affected less by a medicine than will an empty one, and, in narcotic poisoning or in collapse, even an empty stomach may refuse to absorb at all. On the other hand, if the local effect of a remedy be irritating, and if the surface receiving the application be already irritated or inflamed, the local influence will be more intense than ordinary. Constitutionally, too, morbid states may throw out of gear, and most strangely, the usual relation between dose and results. A striking example is in the case of narcotic drugs, which, by the very circumstances calling for their prescription, may require relatively enormous dosage to produce the needed effect—a dosage that ordinarily would even be fatal! Thus in collapse from hæmorrhage a quart of brandy may have to be given to revive the flickering heart, and the effects will be no more than, in health, would follow a tablespoonful. In such cases, therefore, set dosage must be set at naught and the remedy boldly pushed until either the therapeusis sought be obtained or signs of beginning poisoning enforce discontinuance. This effect of morbid status—the *idiosyncrasy of disease*, as it might be termed—is one constantly presenting itself, and one, therefore, that must ever be present in the mind of the prescriber.

Such, then, are, in outline, the considerations affecting dosage, and, reviewing them in thought, an important fact appears, from which follows an equally important corollary: the *fact*, that, concerning a medicine, *dose is not a thing that can be put down at a set figure*, like specific gravity or melting-point, but, being, so to speak, a function of a number of independent considerations, it must perforce vary under varying conditions; and the *corollary*, that, from the very circumstances of the case, *dose is rarely very closely calculable*, the degree of precision attainable, however, differing greatly with different groups of drugs.

APPENDIX.

TABLE OF THE SOLUBILITY OF CHEMICALS IN WATER AND IN ALCOHOL.

[*From the U. S. Pharmacopœia of* 1880.]

Abbreviations: s. = soluble; ins. = insoluble; sp. = sparingly; v. = very; alm. = almost; dec. = decomposed.

CHEMICALS.	WATER.		ALCOHOL.	
	At 15° C. (59° F.).	Boiling.	At 15° C. (59° F.).	Boiling.
One part is soluble in :	*Parts.*	*Parts.*	*Parts.*	*Parts.*
Acidum Arseniosum [1]	30–80	15	sp.	sp.
" Benzoicum	500	15	3	1
" Boricum	25	3	15	5
" Carbolicum	20	—	v. s.	v. s.

[1] Acidum Arsenosum, *U. S. Ph.*, 1890.

TABLE OF THE SOLUBILITY OF CHEMICALS—*Continued.*

CHEMICALS.	WATER.		ALCOHOL.	
	At 15° C. (59° F.).	Boiling.	At 15° C. (59° F.).	Boiling.
One part is soluble in :	*Parts.*	*Parts.*	*Parts.*	*Parts.*
Acidum Chromicum	v. s.	v. s.	dec.	dec.
" Citricum	0.75	0.5	1	0.5
" Gallicum	100	3	4.5	1
" Salicylicum	450	14	2.5	v. s.
" Tannicum	6	v. s.	0.6	v. s.
" Tartaricum	0.7	0.5	2.5	0.2
Alumen	10.5	0.3	ins.	ins.
" Exsiccatum	20	ins.	0.7	ins.
Aluminii Hydras [1]	ins.	ins.	ins.	ins.
" Sulphas [2]	1.2	v. s.	alm. ins.	alm. ins.
Ammonii Benzoas	5	1.2	28	7.6

[1] Alumini Hydras, *U. S. Ph.*, 1890.
[2] Alumini Sulphas, *U. S. Ph.*, 1890.

Ammonii Bromidum	1.5	0.7	150	15
" Carbonas	4	dec.	dec.	dec.
" Chloridum	3	alm. ins.	1.37	alm. ins.
" Iodidum	1	0.5	9	3.7
" Nitras	0.5	v. s.	20	3
" Phosphas	4	ins.	0.5	ins.
" Sulphas	1.3	1	sp.	sp.
" Valerianas	v. s.	v. s.	v. s.	v. s.
Antimonii et Potassii Tartras	17	3	ins.	ins.
" Oxidum	alm. ins.	alm. ins.	ins.	ins.
" Sulphidum	ins.	ins.	ins.	ins.
" Sulphidum Purificatum	ins.	ins.	ins.	ins.
Antimonium Sulphuratum	ins.	ns.	ins.	ins.
Apomorphinæ Hydrochloras	6.8	dec.	50	dec.
Argenti Cyanidum	ins.	ins.	ins.	ins.
" Iodidum	ins.	ins.	ins.	ins.
" Nitras	0.8	0.1	26	5
" " Fusus	0.6	0.5	25	5
" Oxidum	v. sp.	v. sp.	ins.	ins.
Arsenii Iodidum [1]	3.5	dec.	10	dec.
Atropina	600	35	v. s.	v. s.

[1] Arseni Iodidum, *U. S. Ph.*, 1890.

TABLE OF THE SOLUBILITY OF CHEMICALS—*Continued.*

Chemicals.	Water.		Alcohol.	
	At 15° C. (59° F.).	Boiling.	At 15° C. (59° F.).	Boiling.
One part is soluble in :	*Parts.*	*Parts.*	*Parts.*	*Parts.*
Atropinæ Sulphas	0.4	v. s.	6.5	v. s.
Bismuthi Citras	ins.	ins.	ins.	ins.
" et Ammonii Citras	v. s.	v. s.	sp.	sp.
" Subcarbonas	ins.	ins.	ins.	ins.
" Subnitras	ins.	ins.	ins.	ins.
Bromum	33	—	dec.	dec.
Caffeina	75	9 5	35	v. s.
Calcii Bromidum	0.7	v. s.	1	v. s.
" Carbonas Præcipitatus	ins.	ins.	ins.	ins.
" Chloridum	1.5	v. s.	8	1.5
" Hypophosphis	6.8	6	ins.	ins.
" Phosphas Præcipitatus	ins.	ins.	ins.	ins.
Calx	750	1300	ins.	ins.
Camphora Monobromata	alm. ins.	alm. ins.	v. s.	v s.

Cerii Oxalis	ins.	ins.	ins.	ins.
Chloral	v. s.	v. s.	v. s.	v. s.
Chrysarobinum	alm. ins.	alm. ins.	sp.	sp.
Cinchonidinæ Sulphas	100	4	71	12
Cinchonina	alm. ins.	alm. ins.	110	28
Cinchoninæ Sulphas	70	14	6	1.5
Codeina	80	17	v. s.	v. s.
Creta Præparata	ins.	ins.	ins.	ins.
Cupri Acetas	15	5	135	14
" Sulphas	2.6	0.5	ins.	ins.
Elaterinum	ins.	ins.	125	2
Ferri Chloridum	v. s.	v. s.	v. s.	v. s.
" Citras	s.	v. s.	ins.	ins.
" et Ammonii Citras	v. s.	v. s.	ins.	ins.
" " " Sulphas	3	0.8	ins.	ins.
" " " Tartras	v. s.	v. s.	ins.	ins.
" " Potassii Tartras	v. s.	v. s.	ins.	ins.
" " Quininæ Citras	s.	v. s.	ins.	ins.
" " Strychninæ Citras	v. s.	v. s.	ins.	ins.
" Hypophosphis	sp.	sp.	ins.	ins.
" Lactas	40	12	alm. ins.	alm. ins.
" Oxalas	sp.	sp.	ins.	ins.

TABLE OF THE SOLUBILITY OF CHEMICALS—*Continued.*

CHEMICALS.	WATER		ALCOHOL.	
	At 15° C. (59° F.).	Boiling.	At 15° C. (59° F.).	Boiling.
One part is soluble in :	*Parts.*	*Parts.*	*Parts.*	*Parts.*
Ferri Oxidum Hydratum	ins.	ins.	ins.	ins.
" Phosphas [1]	v. s.	v. s.	ins.	ins.
" Pyrophosphas [2]...	v. s.	v. s.	ins.	ins.
" Sulphas	1.8	0.3	ins.	ins.
" Sulphas Præcipitatus [3].........	1.8	0.3	ins.	ins.
" Valerianas.....................	ins.	dec.	v. s.	v. s.
Hydrargyri Chloridum Corrosivum ..	16	2	3	1.2
" " Mite	ins.	ins.	ins.	ins.
" Cyanidum............	12.8	3	15	6
" Iodidum Rubrum	alm. ins.	alm. ins.	130	15
" " Viride [4]	alm. ins.	alm. ins.	ins.	ins.

[1] Ferri Phosphas Solubilis, *U. S. Ph.*, 1890.
[2] Ferri Pyrophosphas Solubilis, *U. S. Ph.*, 1890.
[3] Ferri Sulphas Granulatus, *U. S. Ph.*, 1890.
[4] Hydrargyri Iodidum Flavum, *U. S. Ph.*, 1890.

Hydrargyri Oxidum Flavum	ins.	ins.	ins.	ins.
" " Rubrum	ins.	ins.	ins.	ins.
" Subsulphas Flavus	ins.	ins.	ins.	ins.
" Sulphidum Rubrum	ins.	ins.	ins.	ins.
Hydrargyrum Ammoniatum	ins.	ins.	ins.	ins.
Hyoscyaminæ Sulphas	v. s.	v. s.	v. s.	v. s.
Iodoformum	ins.	ins.	80	15
Iodum	sp.	—	11	—
Lithii Benzoas	4	2.5	12	10
" Bromidum	v. s.	v. s.	v. s.	v. s.
" Carbonas	130	130	ins.	ins.
" Citras	5.5	2.5	sp.	sp.
" Salicylas	v. s.	v. s.	v. s	v. s.
Magnesia	alm. ins.	alm. ins.	ins.	ins.
" Ponderosa	alm. ins.	alm. ins.	ins.	ins.
Magnesii Carbonas	alm. ins.	alm. ins.	ins.	ins.
" Sulphas	0.8	0.15	ins.	ins.
" Sulphis	20	19	ins.	ins.
Mangani Oxidum Nigrum [1]	ins.	ins.	ins.	ins.
" Sulphas	0.7	0.8	ins.	ins.
Morphina	v. sp.	500	100	36

[1] Mangani Dioxidum, *U. S. Ph.*, 1890.

TABLE OF THE SOLUBILITY OF CHEMICALS—*Continued.*

CHEMICALS.	WATER.		ALCOHOL.	
	At 15° C. (59° F.).	Boiling.	At 15° C. (59° F.).	Boiling.
One part is soluble in :	*Parts.*	*Parts.*	*Parts.*	*Parts.*
Morphinæ Acetas	12	1.5	68	14
" Hydrochloras	24	0.5	63	31
" Sulphas	24	0.75	702	144
Phosphorus	ins.	ins.	v. sp.	v. sp.
Physostigminæ Salicylas	130	30	12	v. s.
Picrotoxinum	150	25	10	3
Pilocarpinæ Hydrochloras	v. s.	v. s.	v. s.	v. s.
Piperina[1]	alm. ins.	alm. ins.	30	1
Plumbi Acetas	1.8	0.5	8	1
" Carbonas	ins.	ins.	ins.	ins.
" Iodidum	2000	200	v. sp.	v. sp.
" Nitras	2	0.8	alm. ins.	alm. ins.
" Oxidum	ins.	ins.	ins.	ins.

[1] Piperinum, *U. S. Ph.*, 1890.

Potassa	0.5	v. s.	2	v. s.
Potassii Acetas	0.4	v. s.	2.5	v. s
" Bicarbonas	3.2	dec.	alm. ins.	alm. ins.
" Bichromas	10	1.5	ins.	ins.
" Bitartras	210	15	v. sp.	v sp.
" Bromidum	1.6	1	200	16
" Carbonas	1	0.7	ins.	ins.
" Chloras	16.5	2	v. sp.	v. sp.
" Citras	0.6	v. s.	v. sp.	v. sp.
" Cyanidum	2	1	sp.	sp.
" et Sodii Tartras	2.5	v. s.	alm. ins.	alm. ins.
" Ferrocyanidum	4	2	ins.	ins.
" Hypophosphis	0.6	0.3	7.3	3.6
" Iodidum	0.8	0.5	18	6
" Nitras	4	0.4	alm. ins.	alm. ins.
" Permanganas	20	3	dec.	dec.
" Sulphas	9	4	ins.	ins.
" Sulphis	4	5	sp.	sp.
" Tartras	0.7	0.5	alm. ins.	alm. ins.
Quinidinæ Sulphas	100	7	8	v. s.
Quinina	1600	700	6	2
Quininæ Bisulphas	10	v. s.	32	v. s.

TABLE OF THE SOLUBILITY OF CHEMICALS—*Continued.*

CHEMICALS.	WATER.		ALCOHOL.	
	At 15° C. (59° F.).	Boiling.	At 15° C. (59° F.).	Boiling.
One part is soluble in:	*Parts.*	*Parts.*	*Parts.*	*Parts.*
Quininæ Hydrobromas	16	1	3	1 or less.
" Hydrochloras	34	1	3	v. s.
" Sulphas	740	30	65	3
" Valerianas	100	40	5	1
Saccharum	0.5	0.2	175	28
" Lactis	7	1	ins.	ins.
Salicinum	28	0.7	30	2
Santoninum	alm. ins.	250	40	3
Soda	1.7	0.8	v. s.	v. s.
Sodii Acetas	3	1	30	2
" Arsenias[1]	4	v. s.	v. sp.	60
" Benzoas	1.8	1.3	45	20
" Bicarbonas	12	dec.	ins.	ins.

[1] Sodii Arsenas, *U. S. Ph.*, 1890.

Sodii Bicarbonas Venalis	12	dec.	ins.	ins.
" Bisulphis	4	2	72	49
" Boras	16	0.5	ins.	ins.
" Bromidum	1.2	0.5	13	11
" Carbonas	1.6	0.25	ins.	ins.
" Chloras	1.1	0.5	40	43
" Chloridum	2.8	2.5	alm. ins.	alm. ins.
" Hypophosphis	1	0.12	30	1
" Hyposulphis	1.5	0.5	ins.	ins.
" Iodidum	0.6	0.3	1.8	1.4
" Nitras	1.3	0.6	sp.	40
" Phosphas	6	2	ins.	ins.
" Pyrophosphas	12	1.1	ins.	ins.
" Salicylas	1.5	v. s.	6	v. s.
" Santoninas	3	0.5	12	3.4
" Sulphas	2.8	0.4	ins.	ins.
" Sulphis	4	0.9	sp.	sp.
" Sulphocarbolas	5	0.7	132	10
Strychnina	6700	2500	110	12
Strychninæ Sulphas	10	2	60	2
Sulphur Lotum	ins.	ins.	ins.	ins.
" Præcipitatum	ins.	ins.	ins.	ins.

TABLE OF THE SOLUBILITY OF CHEMICALS—*Continued.*

CHEMICALS.	WATER.		ALCOHOL.	
	At 15° C. (59° F.).	Boiling.	At 15° C. (59° F.).	Boiling.
One part is soluble in :	*Parts.*	*Parts.*	*Parts.*	*Parts.*
Sulphur Sublimatum	ins.	ins.	ins.	ins.
Thymol	1200	900	1	v. s.
Veratrina	v. sp.	v. sp.	3	v. s.
Zinci Acetas	3	1.5	30	3
" Bromidum	v. s.	v. s.	v. s.	v. s.
" Carbonas Præcipitatus	ins.	ins.	ins.	ins.
" Chloridum	v. s.	v. s.	v. s.	v. s.
" Iodidum	v. s.	v. s.	v. s.	v. s.
" Oxidum	ins.	ins.	ins.	ins.
" Phosphidum	ins.	ins.	ins.	ins.
" Sulphas	0.6	0.3	ins.	ins.
" Valerianas	100	—	40	—

INDEX.

PAGE

www.ingramcontent.com/pod-product-compliance
Lightning Source LLC
Chambersburg PA
CBHW021526210326
41599CB00012B/1395
9783337779047